HOW TO SAVE THE AMAZON

DOM PHILLIPS & CONTRIBUTORS

HOW TO SAVE THE AMAZON

A journalist's deadly quest for answers

DOM PHILLIPS & CONTRIBUTORS

ITHAKA

First published in the UK in 2025 by Ithaka Press
An imprint of Bonnier Books UK
5th Floor, HYLO, 103–105 Bunhill Row,
London, EC1Y 8LZ

Owned by Bonnier Books
Sveavägen 56, Stockholm, Sweden

Hardback – 9781786581839
Trade Paperback – 9781786581846
Ebook – 9781786581822
Audiobook – 9781786581679

All rights reserved. No part of the publication may be reproduced, stored in a retrieval system, transmitted or circulated in any form or by any means, electronic, mechanical, photocopying, recording or otherwise, without prior permission in writing of the publisher.

A CIP catalogue of this book is available from the British Library.

Typeset by IDSUK (Data Connection) Ltd
Printed and bound by Clays Ltd, Elcograf S.p.A.

1 3 5 7 9 10 8 6 4 2

Copyright © The Literary Estate of Dominic Phillips, 2025

Dominic Phillips, the Literary Estate thereof, and the contributors have asserted their moral right to be identified as the authors and translators of this Work in accordance with the Copyright, Designs and Patents Act 1988.

Every reasonable effort has been made to trace copyright holders of material reproduced in this book, but if any have been inadvertently overlooked the publishers would be glad to hear from them.

www.bonnierbooks.co.uk

Dedicated to everyone fighting to protect the rainforest

CONTENTS

FOREWORD:
A Bloody Change of Plan — xi

INTRODUCTION: Into the Forest — 1

CHAPTER ONE
Laying Down the Law: Political leadership — 27

CHAPTER TWO
Cattle Chaos: Corporate accountability — 67

CHAPTER THREE
Putting the Eco Back in Economy: Agroforestry models — 109

CHAPTER FOUR
Stop Destructive Development: Managed urbanisation — 149

CHAPTER FIVE
A Cemetery of Trees: Infrastructure catastrophes — 167

CHAPTER SIX
Regrow and Self-Protect: Indigenous defenders — 197

CHAPTER SEVEN
Putting a Price on the Future: Tourism and environmental service payments — 221

CHAPTER EIGHT
Shaking the Global Money Tree: International finance   243

CHAPTER NINE
Nature Worth Fighting For: Biopharmacy and bioeconomy   281

CHAPTER TEN
A Life-Changing Relationship: Educate and rethink   305

AFTERWORD
Listen to the Forest: Indigenous inspiration   341

ACKNOWLEDGEMENTS   355

LIST OF MAPS

Javari Valley	x
The Amazon region	xxxii
Brazil	26
Pará State	66
Amazônas State	108
Belém and Manaus	148
Selected infrastructure of Pará State	166
Roraima State	196
Costa Rica	220
Tapajós Basin	242
Belém and Dubai	242
Maranhão State	280
Acre State	280
Selected mega-projects, completed and under consideration	304
Recognised Indigenous territories	340

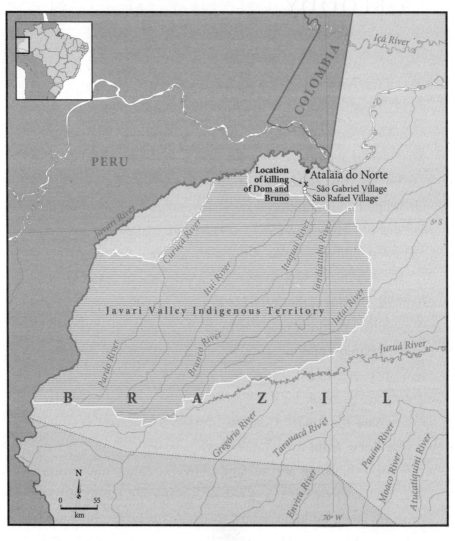

Javari Valley

FOREWORD:
A BLOODY CHANGE OF PLAN

'*Amazônia, sua linda*' (Amazonia, you beauty), Dom Phillips exclaimed in a gloriously euphoric, tragically final social-media post on 30 May 2022. The shout of joy captioned a cellphone video clip of a lush rainforest riverbank, taken from the speedboat that was powering him closer to the Javari Valley – an immense primordial wilderness, one of the last remaining on Earth.

He was on his way to research a story about environmental criminals, the threat they posed to isolated Indigenous peoples and the possible solutions that local campaigners were exploring to resolve a bitter – and increasingly violent – conflict in this border region between Brazil, Peru and Colombia. It was to be a core element in the book he had been working on for several years: *How to Save the Amazon: Ask the People Who Know*.

He knew from a previous expedition in 2018 that the trip would be exhausting and uncomfortable at times, and possibly risky, but for Dom this was a passion as much as an occupation. British by birth, he had adopted Brazil as his homeland and the Amazon as his cause. His last public words convey the joy of a journalist doing a job he loved in a place he cherished.

That night, he made a base in Atalaia do Norte, a run-down frontier town of crumbling concrete sidewalks and a collapsed waterfront esplanade. In the main square, mayors of yore had celebrated the principal activity of the townsfolk, fishing, by erecting concrete likenesses of the giant aquatic beasts caught in local rivers, *pirarucu* (Arapaima gigas), *tucunaré* (peacock bass) and *pirarara* (redtail catfish), which had long since lost their lustre and now resembled gargoyles. The images of the fish were there again in Dom's hotel, the Castro Alves, painted in naif-art murals on the bright mustard-coloured walls and used instead of numbers to distinguish between the second-floor rooms. The ground-floor accommodation was named after Indigenous peoples – Marubo, Kanamari, Matis – who live in the nearby Javari Valley, which is home to 26 ethnic groups, 16 of whom choose to live with as little contact as possible with the outside world.

Fishing and Indigenous culture encapsulate the two conflicting ways of thinking in the region. Outsiders want to extract large quantities of fish for income, for profit, while insiders seek to block them and ensure sustainable catches for their communities. It is a frontline of change, exactly where Dom needed to be for his book, exactly where he hoped to meet people who had optimistic ideas about how to reconcile the many conflicting priorities and motivations in order to save the Amazon. He was determined that his book wouldn't just be about environmental catastrophe; it would be about the people working hard to avert it.

Bruno Pereira, who Dom met up with two days later, on 1 June, was chief among them. Described by friends as a force of nature, Bruno was the co-founder of a group of Indigenous forest guardians who patrolled the borders of the Javari Valley and

Foreword: A Bloody Change of Plan

collected evidence of crimes using drones, camera traps and GPS trackers. He knew the territory and people intimately, could speak and sing in ethnic languages, and had the reputation of being fearless in the face of death threats – of which he had received many.

Four years earlier, the burly and brilliant *indigenista* (Indigenous specialist) had led Dom's first expedition into the Javari Valley. After that, he became an inspiration to Dom and a mentor. Without his encouragement, Dom would not have dreamed of writing a book on such an ambitious theme.

Bruno had invited him back to Javari because an already complex situation was becoming tenser. The Amazon was being taken over by narco-traffickers. Organised crime was moving into the region, using the nominally closed-to-outsiders Javari Valley as a route for smuggling drugs and guns. This invasion by armed gangs posed a dire threat to the Indigenous people in the forest, but the Brazilian state, at that point headed by the far-right former military captain Jair Bolsonaro, who had become president in January 2019, was doing nothing. In fact, worse than nothing. Bolsonaro was encouraging invasions of the forest by illegal miners, land-grabbers and loggers, thereby raising the risks for those who defend it. In reaction to Bolsonaro's presidency, Bruno had taken a leave of absence from his government work with the National Foundation of Indigenous Peoples – the main state organisation responsible for the welfare and territories of Brazil's original population, usually known by its Portuguese acronym FUNAI – to join forces with an association of local Indigenous people. Together, they had established the Javari Valley Indigenous Association Patrol Team.

Dom needed little persuading to return to Javari, though he was aware of the risks and discussed them with his Brazilian wife, Alê (Alessandra Sampaio): 'Dom knew about the threats that Bruno faced. But he said to me, "Bruno has faced threats for ten years. He is aware. He is alert." But we knew it was not completely safe.'

The tension, however, was worse than he anticipated. The night before he was due to set off with Bruno on a boat up the Itaquaí river towards the Javari Valley, Dom interviewed two other members of the Indigenous patrol team, known by its Portuguese initials EVU, who told him how bad things had become. The veteran Cristóvão Negreiros Pissango, known as Tataco, told him the risks had increased since Dom's last visit, four years earlier. He and Bruno had been shot at the previous month by an illegal fisher called Amarildo. A year or so before, Tataco had been punched in the face by another fisher in the town square. Several years earlier, when Tataco was working for FUNAI, one of his colleagues, Maxciel Pereira dos Santos, had been murdered. 'This is a war. So many people are dying,' he said, and offered to join the following day's expedition to provide extra protection.

Dom's unease intensified when he talked to Orlando Possuelo, who was a co-founder of the EVU. He had received death threats in front of his family and warned Dom that Bruno was also a target. He too offered extra security for the trip. Bruno brushed off the offers and the fears, joking to Orlando, 'You have made Dom scared.' He would take him alone.

Bruno had grown used to living under a near constant state of threat. He often travelled the river with just one or two others. Plus, foreign journalists had been coming and going to the Amazon for decades without major incidents – though Javari was admittedly one of the more out-of-the-way destinations.

Foreword: A Bloody Change of Plan

Bruno could not have anticipated the extraordinarily awful events that would follow.

The last known photograph of Dom is a source of both solace and horror to those who knew and loved him. The image was taken by Bruno at 7.03am on the morning of 5 June 2022 in the fishing village of São Gabriel on the Itaquaí river, close to the gateway of the Javari Valley. Dom is wearing a dark yellow shirt and sitting on a boat, listening intently to a man holding a child, from one of the dozen or so families in this small, poor community.

This picture is consoling because it confirms Dom was doing what gave him most satisfaction, right up until the last few minutes of his life. There he is, all ears, captured in the moment, trying to understand a complex, important conflict by asking those who knew it best. The previous day, he had interviewed the Indigenous protectors of the rainforest. Now, he was talking to the invaders, the riverine villagers who illegally entered Javari to fish and poach. In an area of considerable conflict, Dom was calmly and bravely living up to his journalistic ideals, listening to both sides of the story.

Yet that is also what is so chilling about this image. In retrospect, we wish this interview had never happened, that Dom and Bruno were anywhere but here, on this part of the river at this particular time, because we know that within an hour, they would both be dead. Police believe they were betrayed by the man in the photo, Janio Freitas, from the village of São Rafael. As soon as the interview was over, he allegedly radioed ahead to the next village, just a few bends of the river downstream, to alert them that Bruno was heading their way.

Lying in wait there were local fishers Jefferson da Silva Lima and Amarildo da Costa de Oliveira, known locally as Pelado. Amarildo was the man who had fired a shot at Bruno a month earlier, as Tataco had told Dom. He had a longstanding grudge against Bruno, who he blamed for past confiscations of his illegal catch and for training the Indigenous patrols that made his life difficult. Amarildo had sworn to kill him and he now had the opportunity to do so.

Two months earlier, police say, he had bought a new 60-horsepower speedboat engine with money from a wealthy local mobster known as 'Colombia', who was actually from Peru. This meant he had one of the fastest vessels on the river, with 50 per cent more power than the boat piloted by Bruno. Once Amarildo started to chase, there was no way Bruno and Dom could escape. The only consolation, according to Bruno's former colleagues, is that the motor on their own craft was so noisy that they were unlikely to have been aware they were being hunted down until the last moment, sparing them minutes of fear.

In testimony to police, Amarildo and Jefferson said they fired at Bruno, who was hit by three shots in the back. A giant of a man, Bruno did not immediately succumb. He too was armed and fired off several wild shots before he collapsed on the controls, skidding the boat towards the riverbank, where it crashed through the undergrowth, snagging on branches and tendrils, before jolting to a halt. Amarildo then moved in for the kill, finishing off Bruno and finally Dom, who put his hands up and implored 'no' before a fatal blast to the chest.

Like many journalists before and since, Dom was killed because a criminal needed to silence a witness. But how had he ended up

Foreword: A Bloody Change of Plan

risking his life in the Javari Valley? And what had made him set out to write such an ambitious and, in the end, dangerous book?

There was little in the first 50 years of his life to suggest that Dom would die such a death in such a place on such a story. A decade earlier, he was freelancing for an oil and gas industry newswire. Ten years before that, he had been leading a safe, hedonistic life as one of Britain's leading music editors. As a child, he had believed that the cowboys were the good guys and the Indians the enemies. Why had he been willing to risk his life to raise awareness of the solutions to the Amazon's problems being proposed by Indigenous people and forest guardians?

Dominic Mark Phillips was born on 23 July 1964 in Collingwood Road in Bebington, a leafy town on the Wirral, the finger of land on the other side of the Mersey river from Liverpool in northwest England. Later in life, his accent would prompt some to call him a scouser – the nickname for people from Liverpool – but Dom would always correct them, pointing out he was a 'woolly back' from the sticks. Not that he was particularly proud of his suburban, middle-class origins, he just wanted to be clear and exact.

His roots were Irish on his father's side and Welsh on his mother's. These areas of the British Isles were holdouts of the original peoples who had been pushed to the geographic fringes more than a thousand years earlier by successive invasions of Anglo-Saxons and Normans. And like Indigenous peoples in other colonised countries, their people were historically treated as second-class citizens by the settlers and their descendants.

Dom's paternal grandfather had been a carpenter, who lived in a two-up, two-down terraced house in the working-class neighbourhood of Rock Ferry, Birkenhead. Postwar opportunities in education opened the door to greater social mobility for the next generation. His son Bernard (Dom's father) won a scholarship to St Anselm's College, a Catholic school run by the Christian Brothers from Ireland, then became a chartered accountant and later a lecturer at Liverpool Polytechnic. Dom's mother, Gillian, qualified as a school teacher after raising the couple's three children: Dom and his younger twin siblings, Sian and Gareth.

Like his father, Dom passed the 11-plus exam and won a scholarship to St Anselm's grammar school. Although he enjoyed watching football and supported the local team, Everton, Dom hated playing rugby and athletics due to the chronic asthma that had required several hospitalisations during his childhood.

His academic record was mixed. Dom's greatest influence at the school was a young English teacher, Jean Parker – one of the few women in the building – who recommended books and music outside the regular coursework. In other subjects, Dom's attention sometimes drifted and he was punished with the strap by the Brothers. His final report card, still kept on file by the school, shows he passed his English language O-level a year early in 1979 and then added seven other O-levels the following year. In his A-level exams in 1982, he earned four grade Ds and below, which was not good enough for the university place he hoped for. By then, though, he had other priorities.

Music had more or less taken over his life. In Dom's early teens, he had started spending most of his pocket money and paper round income on records. There was a punk phase, when he listened to The Clash, The Stranglers, The Slits and Elvis Costello, followed by New Wave. As soon as he got home from school, he would throw off his royal blue blazer and dress up in the trendy

Foreword: A Bloody Change of Plan

eighties style, with a long woollen coat from a charity shop and a trilby hat, then head out to the local pub, the Rose and Crown, until he got thrown out for underage drinking.

On weekends, he would travel hours to see gigs in Liverpool clubs and as far away as Wrexham. He formed a garage band, SPK, in which he played bass guitar. Sometimes he would jam with his brother Gareth, who was on drums. The neighbours were not impressed by their covers of The Doors and Velvet Underground and Dom's original compositions, but the band reached a high enough standard to play the school disco, a club in Birkenhead and, most impressively, Brady's in Liverpool, which was where U2, Depeche Mode and Duran Duran had cut their teeth.

Family disaster then turned his plans upside down. When Dom was 19, his father had a heart attack and slipped into a coma. Starved of oxygen for more than ten minutes, Bernard's brain was damaged beyond any hope of recovery. He spent the rest of his life – more than ten years – in a hospital in Wales, which had a knock-on effect on his devoted wife, Gillian. 'Mum sold our home and moved there so she could visit him. That became her life,' Sian recalled. 'It was a very hard time. A really tragic story.'

Dom was unsettled. After brief, unhappy stints as a literature student at Hull University and Middlesex Polytechnic, he dropped out and started travelling. He busked his way around the Mediterranean, then lived for several years in Denmark before returning to the UK and re-immersing himself in the music scene. He lived mainly in squats and hustled freelance music articles wherever he could. He knew he wanted to be a writer.

In Liverpool, Dom started a music fanzine named *The Subterranean*, after Jack Kerouac's novel, with a friend who worked for the civil service and, crucially, had access to a printer. In 1988, he launched another fanzine, *New City Press*, in Bristol and started a weekly radio show under the name DJ Banjo with his co-founder John Mitchell, a.k.a. DJ Yogi. Among his many ventures in this period were in-depth features on the Barton Hill Youth Club, where graffiti artists such as Banksy honed their craft, and the co-curation of a charity album, *The Hard Sell*, which included a first solo outing from Tricky of Massive Attack. In 1991, Dom finally got a break when he was appointed production manager for a tiny magazine called *Mixmag* that was about to become the bible for a new generation of ecstasy-fuelled acid house ravers. Dom moved to London and drove every day down the M4 to *Mixmag*'s offices in Slough.

The timing could not have been more fortuitous. Underground acid house raves exploded into the global consciousness in the form of a lucrative dance music scene. The 1990s was the decade of the DJ and Dom knew them all, as 'work' took him from superclubs in London, like the Ministry of Sound, to smaller, but often livelier, franchises in the north of England, where he was convinced the best venues were to be found. Interviews with Björk and the DJs Sasha, Pete Tong, Jeremy Healy, Dave Seaman, Nicky Holloway and Fatboy Slim led to a surge in readership.

The magazine fronted the wild rollercoaster of dance music as it inspired a nation of young people to come together and have a good time. Initially, this was not just about beats per second and psychedelic trances: at its heart was a subculture that promised – in that early phase at least – peaceful, interracial, revolutionary change and a challenge to the commercialisation of music.

In the following years, the scene grew far beyond these idealistic roots into an era of chemically fuelled energy and attitude.

Foreword: A Bloody Change of Plan

For the DJs and the club owners, these were heady times. As Dom would later write: 'It was about larging it. It was about pulling out a wad of 20s when you were buying your champagne at the bar. It was about buying your cocaine in an eight ball. It was about wearing designer clothes. At that top tier of that club scene, it was about giving it loads.'

Style was everything. 'Nothing feels better than a new haircut,' Dom insisted, and he lived up to this mantra, always immaculately trimmed and turned out in the latest fashions. Now one of the country's most influential music journalists, he was such a culture aficionado and dedicated follower of fashion that friends nicknamed him 'Mr Cool'. That was the ultimate aspiration for many Brits of his generation, but he came to crave something more.

By 1997, Dom was editor-in-chief of *Mixmag*, which was then selling over a hundred thousand copies a month. He broke countless old-fashioned journalistic taboos, putting an ecstasy pill on the cover of one issue, a cow in a field on another, then police chasing rioters in Trafalgar Square. Features were equally wild and exciting. Would you let this horse into your club? What's it like to drive on drugs? Was gangland invading clubland? He followed DJs around the world to clubs in Europe, North and South America and Asia, where they could command £140,000 a night. The reporting was no-nonsense and first person.

The result was ever-increasing sales and in 1997, the magazine was acquired by EMAP, a pushy and fast-growing publishing house. Dom barely lasted a year. The early idealism had been bought out like everything else. Corporate life was not for him and he couldn't help noticing that many of the black contributors had stopped coming to the office.

He grew disillusioned, lamenting later that the optimism and excitement of the early club scene had been replaced by smoke and mirrors. 'These were the boom years of fast living and easy money. Acid House was unstoppable. And then it all went spectacularly wrong,' he would later write. 'Dancing wildly in nightclubs, as anybody who spent time doing so in the 1990s knows, is an experience as intense as it is inane, as emotional as it is ephemeral.' Underneath the shiny surface, he observed, 'lurked a darker side, a world of cynical money making, rampant egos and cocaine-fuelled self-indulgence that eventually spiralled out of control leaving behind burnt-out DJs, jobless promoters and a host of bittersweet memories.'

The comedown coincided with personal turmoil. His father and mother died within a month of each other in 2000. His marriage with his first wife Nuala started falling apart and ended several years later in an acrimonious and costly divorce. But rather than dwell on his misfortunes, Dom decided to write up the rise and fall of the club scene in the form of a book. He secured an advance that was respectable but would not quite fund a London lifestyle. He was restless, he needed to get away from soured relationships, he wanted to rediscover a feeling of joy. In his mind, there was only one place to go for a new start in life.

Mixmag had taken Dom up and down the motorways of Britain to pretty much any decent club the country had to offer. It had also taken him around the world – New York, Paris, Singapore and beyond. Among all these destinations it was Brazil with which he felt an affinity. He had made friends there and started returning whenever he could raise the funds or secure a writing commission. He had done the maths: the publisher's advance would just about

Foreword: A Bloody Change of Plan

cover his costs to live in São Paulo and write the book, which now had a title: *Superstar DJs Here We Go! The Rise and Fall of the Superstar DJ*. So off he flew. It was 2007. He never moved back.

Once the book was published, Dom was free to delve deep into the diverse culture and belief systems of his adopted homeland. He worked hard to learn Portuguese and steadily picked up writing gigs. When he moved from São Paulo to Rio de Janeiro, he started to feel that he wanted to put down roots. On 2 February 2013, he celebrated the Candomblé festival of Yemanjá Day with a supplication to the spirits to find him a 'nice girlfriend'.

Twenty days later, at a friend's party in Rio's bohemian district of Santa Teresa, he met a smart, kind and beautiful woman from Salvador by the name of Alessandra Sampaio. They talked and talked about art, music and the fabulous city they both now called home. 'I looked at him and thought, what an amazingly interesting guy,' she remembers. Alê had previously studied in London but she was too shy to speak in English, so they chatted in Portuguese, which Dom had quickly mastered. They went their separate ways at the end of the night, but the spirits were not to be so easily denied.

To their mutual surprise and delight, Dom and Alê bumped into each other on Ipanema Beach the following day. A week later, they started dating and from then on their lives were intertwined. They moved in together the following year and married two years later. The reception on 5 December 2015 was a joyously raucous affair in Santa Teresa, celebrated by friends from the different phases of Dom's life – British clubbers, Brazilian artists, foreign correspondents and his new family in Brazil. The dancing went on beyond dawn.

HOW TO SAVE THE AMAZON

Back then, Dom was still finding his feet as a journalist, freelancing for a range of publications including *Folha de S.Paulo*, *The Times* of London, the *Financial Times*, *FourFourTwo* (a football fanzine) and *Platts* (an energy and industry newswire). 'He was working for an oil and gas magazine. He didn't like it,' Alê said.

So he turned to other work, with more of an environmental bent, for the *Washington Post* and *The Guardian*. 'He told me he wasn't a trained journalist because he had never gone to university, but he was determined to make it as a foreign correspondent. It was inspiring. He worked so hard. Sometimes he had to write and rewrite four times to satisfy the editors. But each time he learned and burnished his skill as a journalist,' Alê remembers.

Dom learned fast, was generous with his time, and became a highly respected and well-liked member of the informal band of foreign correspondents in Brazil. His growing reach took him eventually to the Amazon, to Bruno and to the Indigenous people of the Javari Valley. After that, no other subject interested him more.

His concern for the region was heightened at the end of 2018, when far-right former military captain Jair Bolsonaro won the Brazilian presidential election on a pledge to halt demarcations of Indigenous land, weaken environmental groups and encourage extraction of natural resources in the Amazon.

Dom shared his fears in a WhatsApp message after the first round of voting: 'This is a very dark and worrying period and it's only going to get worse,' he wrote. 'My sense is that it is also going to become more dangerous for journalists.' His greater concern was for defenders in places like the Javari Valley. Dom was sure a second-round victory for Bolsonaro would give thugs a green light to step up their assault. 'If he wins, what will living

Foreword: A Bloody Change of Plan

here be like? It's like carte blanche to attack anyone his mob disagrees with,' he warned.

The new president proved every bit as bad for the forest as predicted. He slashed the budgets for government protection agencies, put political appointees in charge of running them down, verbally attacked environmental NGOs and gave a green light to illegal miners, loggers and land grabbers. In the Javari Valley, the local FUNAI office was put under the control of an evangelical zealot whose primary goal was to open up the area to missionaries. The government had effectively switched sides: instead of protecting the rainforest, it was now an additional threat.

It was in this context that Dom decided to write *How to Save the Amazon*. It was one of two big projects Dom had underway at the time, along with a plan for him and Alê to adopt a child. Dom had secured a publishing contract with Bonnier Books UK. He promised them 'a character-driven, deeply researched, campaigning, environmental travel book that aims to entertain, inform and, most importantly, mobilise readers. I want people to think differently about the world's greatest rainforest and how they can contribute towards protecting it by taking them there to meet Indigenous tribes, forest peoples, social and business entrepreneurs, environmentalists, scientists, economists, anthropologists and farmers, all of whom know and understand the Amazon intimately and have innovative solutions for the millions of people who live there.'

Despite the uncertain financial situation that he and Alê were in, he took a year off from his usual freelance work so he could devote himself to this hugely ambitious topic. He quickly burned through his advance on reporting trips to the Brazilian Amazon and Costa Rica (where he aimed to look at alternative solutions). A grant from the Alicia Patterson Foundation provided some breathing space, but the Covid pandemic hampered his progress and left him short of time as well as cash. To save money, he and Alê gave up their flat in Rio de Janeiro and moved to a cheaper place in Salvador, Bahia. Even this was not enough. Dom had to ask his family back in England for a loan so he could make one or two more field visits and have enough to live on while he finished the book.

He never got the chance. Alê's final conversation with her husband was on 2 June, when he was in Atalaia do Norte. Dom said he had met Bruno, shared a taxi with an Indigenous family and interviewed the son of Sydney Possuelo, the most famous Indigenous expert in Brazil. A few hours later, he recorded a voice message, passing on all his contacts and the schedule for his forthcoming river trip to the edge of the Javari Valley. He promised he would call her when he was back in cellphone range on Sunday the fifth or Monday the sixth at the latest. She remembers that he signed off: 'Love you. Miss you.'

On Monday 5 June, Alê received a call from one of Dom's friends to say her husband was missing. He and Bruno had not arrived back in Atalaia do Norte as expected the previous day. Possuelo and the Association of Indigenous Peoples of the Javari Valley had scoured the riverbanks in vain. She immediately tried to call Dom and left messages, but feared in her heart that 'it was too late; he was already in another dimension.'

Soon after, word started to spread more widely. A WhatsApp group of Dom's journalist friends badgered every editor, influencer and celebrity they could think of to encourage the Brazilian

Foreword: A Bloody Change of Plan

government to step up the search. Their past experience of missing persons – not just in Brazil, but around the world – suggested the authorities would do the minimum unless there was a public outcry.

Twitter rapidly buzzed with thousands of appeals for the government to mobilise more personnel to find the men. There were supportive tweets from the former Brazilian president Luiz Inácio Lula da Silva, Seleção footballer Richarlison and Hollywood actor Mark Ruffalo. Brazilian music legend Caetano Veloso echoed concerns during his shows, while the designer Cristiano Siqueira created striking black-and-white portraits of the two men on a red background emblazoned with the question 'Where are Dom Phillips and Bruno Pereira?' This image quickly became ubiquitous, appearing on city walls around Brazil, flashing out from advertising trucks in Los Angeles and beamed onto buildings in London.

The disappearances made headlines across much of the globe, were discussed in the British Parliament and raised by US congressmen. Brazilian president Jair Bolsonaro had to respond. It took several days before he mentioned the case of the two men, who he claimed had embarked on an inadvisable 'adventure', implying they were somehow to blame.

Meanwhile, in the Javari Valley, the Indigenous surveillance team continued the search and their indefatigable efforts turned up key pieces of evidence, including the poignant discovery of Dom's National Union of Journalists press card. The worst fears of Alê and Bruno's wife Beatriz Matos were confirmed on 15 June, when the bodies were found. Amarildo, his brother Oseney and Jefferson were charged with murder and disposing of the body.

HOW TO SAVE THE AMAZON

The suspected mastermind Colombia was arrested for document fraud and questioned about where he got the money to bankroll the illegal fishing gangs that plunder the Javari Indigenous territory. Two local politicians were put under investigation for their possible connection to Colombia's alleged crimes. At the time of going to press, all these cases were ongoing.

The repercussions were enormous. Grief and anger spread through the Javari Valley and across the world. The Kanamari mourned for a whole year. In line with their mourning custom, they shaved their hair, forbade certain crops, and rained songs and tears on the memory of the deceased. The Union of Indigenous Peoples of the Javari Valley (Univaja) changed its logo to a black ribbon until the first anniversary of the death, after which they erected two giant wooden crosses on the forested bank of the Itaquaí river, overlooking the spot where the two men were killed. The nearby scene was very much like that in Dom's final '*Amâzonia, sua linda*' social media post.

Dom and Bruno's images, seemingly locked side by side in eternal brotherhood, would also adorn carnival floats in Rio de Janeiro, be the subject of an art installation at the São Paulo Biennale, inspire at least four film documentaries and become icons for environmental protection and press freedom. The two men had joined the ranks of the Amazon forest martyrs, alongside other murdered environmental heroes such as Chico Mendes, the rubber tapper and trade union leader who was killed in 1988 in Xapuri, Acre, by a rancher for trying to preserve the Amazon rainforest and the rights of its people; Dorothy Stang, the American nun who was assassinated in Anapu, Pará State, in 2005 for helping smallholders to secure land rights in defiance of threats from powerful

Foreword: A Bloody Change of Plan

landowners, and hundreds of other Indigenous and forest activists whose killings largely go unreported and uninvestigated.

As the justice system crawled forward, memorial services were held for Dom and Bruno in Rio, São Paulo and London. For many of his friends, the whole thing felt surreal. It was hard to accept the utter horror of what had happened. Yet it was also important to acknowledge that killings of land defenders and environmental reporters were an all-too-common occurrence both in Brazil and around the world. The main reason this case had a high profile was because one of the victims was a white, foreign journalist.

The collaboration for this book emerged from those dark days. Finding a way to finish Dom's work was initially a way to deal with grief for those close to him – his family, friends and colleagues. Coming together on this project enabled us to share our sadness, honour Dom and look forward. In doing so, we experienced for ourselves the true meaning of solidarity. Now we had a purpose. Nothing good could come from such a heinous murder, but we could at least prevent the killers from silencing the story that Dom had been trying to tell.

How to Save the Amazon became a rallying point. Like the murals of Bruno and Dom that sprang up on walls around Brazil, the Rio carnival floats bearing their images, the T-shirts with their faces, or the lapel badges bearing the message '*Amâzonia, sua linda!*', it was a way to keep their memories and their work alive.

Alê, to whom Dom had left his literary estate in his will, appointed the small coordinating committee that has put together this foreword. We, in turn, selected contributors for each chapter. Dozens of other prominent journalists volunteered to proofread and edit. Dom's family raised awareness of the book and helped fundraise. Hundreds more friends and well-wishers participated through donations to the fundraising campaigns to pay for reporting trips, translators and fact-checkers. Thousands spread the word on social networks.

At the time of Dom's death, less than half of the book was finished. After writing the detailed proposal on which Bonnier had acquired the book, he had drafted an introduction, the first three and a half chapters, and left notes, transcripts and plans – in sharply varying degrees of detail – for the remaining six. Picking up where he left off digitally was relatively easy as his files were backed up. Following where he was going in terms of the story was much tougher because Dom was an old-school journalist who wrote everything down in ring-binders and deciphering his scrawled notes proved far trickier than cracking his computer encryption.

Chapter contributors were asked to follow Dom's plans as closely as possible, to retrace some of his steps, to interview people he had talked to, and to try to find and assess the solutions he was looking for. We asked them to make each chapter a dialogue with Dom, through his notes and conversations. In some cases, this was relatively straightforward. The chapter on Costa Rica, for example, was well advanced and the writer, Stuart Grudgings, lives in that country and was already familiar with Dom's ideas for this part of the book. In other sections, Dom was still at an early stage in his thinking, so contributors for sections on international finance, biotechnology, culture, the media and social networks had much less to go on, and were advised instead to engage in an inner dialogue with what they knew of Dom and to focus – as he did throughout – on the search for remedies. In that sense, these chapters are an attempt to unravel a riddle and a hunt for clues to

Foreword: A Bloody Change of Plan

try to trace what Dom had been looking for. Like Dom, none of us was under any illusion that our writing would save the Amazon, but we could certainly follow his lead in asking the people who might know.

We hewed as closely as possible to the spirit and letter of Dom's intentions, with a couple of exceptions. We had to adapt because the situation in Brazil had changed, in part due to the 2022 presidential election victory of Luiz Inácio Lula da Silva, who ushered in a more proactive policy of defending the Amazon rainforest. And also because, whether we liked it or not, the deaths of Dom and Bruno were now part of the story. We reasoned that Dom would also have wanted to update his material (he mentioned the forthcoming election of October 2022 in his proposal and the fact that he would need to take its result into account) and that even though he felt congenitally uncomfortable about journalists writing about journalists, he might make an exception in this extreme case.

No reader should be in any doubt that these pages have been stained by blood. The killers blasted a gaping wound in this book that is far too great for any infusion of solidarity to heal. We hope that what is lost in the clarity of a single voice is made up for by a diversity of perspectives and styles. The writers may have a different take, but all are aligned in ensuring Dom's work is completed and able to live on long after him. This is still very much his book. Dom's, Bruno's and the Amazon's.

Rebecca Carter, David Davies, Andrew Fishman,
Tom Hennigan and Jonathan Watts
January 2025

The Amazon region

INTRODUCTION: INTO THE FOREST

Dom Phillips
The Javari Valley, Amazonas State

'The precious Amazon is teetering on the edge of functional destruction and, with it, so are we.'

'SNAKE!'

The cry came from near the end of the file of eleven men, strung out along a narrow trail being hacked out of thick Amazon rainforest. I shivered. I had walked right past the danger lurking unseen in the dense undergrowth. Poisonous snakes are one of the most lethal threats in this part of the world. Indigenous people fear them and they present even more danger to a bumbling, middle-aged journalist like me, stumbling over roots the local men stepped lightly over in their rubber boots, skidding on muddy ground where they were surefooted.

Takvan Korubo – a taut, forbidding man with a wicked sense of humour – was unfazed. He knew these forests like city people know streets. Snakes for him were an occupational hazard, an everyday danger. He shouldered the polished, shoulder-height wooden club his Korubo people use for hunting and fighting, and walked briskly back towards the reptile with all the care of a man heading down a garden path to fix the gate. 'Be careful,' shouted Bruno Pereira – an official from FUNAI, the Brazilian government Indigenous agency, who was leading this expedition.

HOW TO SAVE THE AMAZON

There was a short, terse silence. Three loud thumps. Seatvo, a Korubo boy of around 13, appeared grinning hugely, dangling the thick body of a one-and-a-half metre, poisonous jararaca on a stick. 'If this bites you, you won't live,' said Josimar Marubo, another of the Indigenous villagers on the trip. He told the grisly tale of a local woman bitten in the breast by a similar-sized jararaca. Not even the serum Pereira was carrying could save her.

The snake was an unsettling reminder of the dangers of the wilderness we were in: the Javari Valley, a vast, inaccessible Indigenous territory of sinuous rivers and dense forests on the Western fringes of the Brazilian Amazon. Apart from the giant hornets, the caiman, the anacondas, the jaguars and the electric eels we had been gleefully warned about, there were snakes everywhere. The men killed a small jararaca when we sat by a creek for lunch, tossing its limp body aside to munch biscuits, then killed another little jararaca while clearing undergrowth to break camp.

I wasn't thinking about how the Amazon might be saved when I set out for the Javari Valley in 2018. I was thinking about how it was being destroyed. I had been living in Brazil for over a decade and was increasingly drawn to stories from the Amazon – a vast basin, twice the size of India, that surrounds the Amazon river and encompasses swathes of Brazil, Peru, Bolivia, Ecuador, Colombia, Venezuela, Guyana and Suriname. Until 50 years ago, it was largely rainforest, but that was changing with horrifying speed. I had first visited in 2004 on a holiday to the sweltering city of Belém with its market full of bizarre-looking river fish. I took a boat to the island of Marajó on the mouth of the River Amazon, where police rode buffalos, the tree wall was resplendent in a thousand shades of green, vultures picked the dried-out

Introduction: Into The Forest

carcass of a cow on a dirt road and the sunsets were a cinematic explosion of pink, turquoise and orange. It wasn't my first trip to Brazil but it was the one that finally hooked me. I moved to São Paulo three years later.

The Javari trip was an assignment for *The Guardian* newspaper. Photographer Gary Calton and I joined an expedition Bruno Pereira was leading following an invitation from the valley's Indigenous association, UNIVAJA, who wanted the dangers the reserve faced exposed. Pereira was looking for signs of a voluntarily isolated Indigenous group in order to ensure its protection.

Designated a protected Indigenous territory in 2001, the Javari Valley is home to about 6,000 Indigenous residents from seven peoples, who share it with at least 16 voluntarily isolated Indigenous groups, a higher concentration than anywhere else in the world. In 2018, the land inhabited by the *isolados*, as these isolated groups are known, was more threatened than it had been in decades – by contamination from heavy pollutants, illegal gold mining barges entering its rivers to the east, armed commercial fishing gangs prowling and cattle ranchers pressuring its southern edges. We overtook one of the fishing gangs' wooden boats, towing canoes behind it, just outside the reserve. A bare-chested man standing on its roof watched as we cruised past, his face creased with suspicion at the green FUNAI uniforms of the government Indigenous agency that Pereira and his team, along with Gary and me, were wearing.

The burly, bearded and bespectacled Pereira, a serious and committed public servant, woke up at 3.30am every morning at camp to plan the day's route. He slept without a mosquito net and eschewed repellent, even though he had caught malaria several times. One morning, as he nonchalantly ate monkey brain for breakfast with a spoon while wearing shorts and flip flops, he discussed government policy towards Brazil's 1.7 million Indigenous people. He had spent years working in the region and was an expert on its voluntarily isolated groups. Before we set off, he had explained that some Marubo people living in a hamlet called São Joaquim deep inside the Javari Valley had been unnerved by fleeting visits from naked, long-haired *isolados*. The plan was to look for signs and clues that they might have moved nearer to São Joaquim.

There are around 100 of these isolated peoples in Brazil. 'Isolated' is not completely adequate as a description because many are believed to have fled enslavement and murder decades earlier, which is why 'voluntarily isolated' is more widely used. Many of these people know outsiders are there but choose to live outside modern technology and Brazilian society. Many are nomadic or semi-nomadic hunter-gatherers, but also manage small plantations cleared out of the forest, and are highly vulnerable to even simple diseases like flu. Monitoring them involves overflights, intelligence and gruelling, often dangerous missions like this, where FUNAI looks for signs but avoids contact, a policy it has held since 1987.

'It's not about us,' said Pereira. 'The Indigenous are the heroes.'

He was the only FUNAI employee on the expedition. Marcir Ferreira, a fisherman and backwoodsman who also worked as a boatman for the mayor of Atalaia do Norte, a town just outside Javari's northeast border, had been contracted to come along. So had Daniel Mayoruna, an Indigenous man from the reserve's Mayoruna people with extensive experience of expeditions like this. There were also two men from the Marubo people, Alcino and Josimar, both of whom had been on similar expeditions before.

Introduction: Into The Forest

The Marubo is the largest of Javari's peoples. First contacted over a century ago, they are regarded as the valley's diplomats, as they often serve as interlocutors with Brazilian and foreign authorities, and live in elaborately constructed, high-ceilinged, thatched communal huts called *malocas*. Josimar lived in the tiny São Joaquim hamlet where the *isolados* had been spotted on several occasions in recent years, stealing bananas, axes and machetes, and even leaving a freshly killed *cutia* – an agouti, which resembles a giant guinea pig – as a present.

Pereira had also invited four Korubo Indigenous people on their first expedition: Takvan (the adept snake killer), his adopted son Xikxuvo Vakwë, Lëyu and his son Seatvo. There are around 100 Korubo, and they are a warlike people known as the *caceteiros* – or 'clubbers' – for the wooden clubs they carry. As well as defending them against snakes, these clubs are also used against those they feel threatened by.

The first group of Korubo was contacted in 1996. This people were believed to have killed a FUNAI employee a year later. A second group became embroiled in a bloody contact with the Matis, another valley people, in 2014 that left dead on both sides before Pereira and FUNAI stepped in to negotiate a fraught contract. Xikxuvo Vakwë was adopted by Takvan at the same time. He said his father had been poisoned by his own brother, covetous of his wife. The group was gripped by an epidemiological crisis. 'There was not much food,' Xikxuvo said. 'There was a lot of fever, headaches.'

We stopped by a Korubo village on the way upriver, where Pereira bantered with a young man who spoke good Portuguese. The Korubo villagers sat on logs under a thatched roof as children and tiny pet monkeys played in the dust. They were naked, coated in the red dye of urucum seeds, or occasionally wearing a pair of shorts or football shirt, with their bowl-shaped haircuts shaved at the back of the head. They told us that four fishermen from one of the fishing gangs had fired over the heads of a group of children that morning. Invasions from fishing gangs were on the rise and they were deeply concerned. Their huts, made of lattices stretched around wooden beams and covered with palm-fronds, were smaller and less sophisticated than the Marubo's communal houses. The Korubo woke before dawn, chattering noisily, and were so expert at imitating the calls of birds and monkeys that the animals replied. They tied the heads of their penises to their waists with twine. At night, Takvan and Lëyu often stripped off their FUNAI gear to relax naked around the fire. Of the two Indigenous peoples, the Korubo seemed especially at ease in the trees. 'They walk in the forest more than us,' said Josimar Marubo, a quiet, capable man with an easy smile.

Every step of the journey presented obstacles, none of which appeared to worry the men. When our boat journey up the Sapóta River from Josimar's village of São Joaquim was blocked by a huge fallen tree, the men spent 90 minutes hacking at it with an axe. Then, when the trunk was almost cut through, they jumped together along it, laughing uproariously when it finally shattered and flung them into the water. They leapt into the river waters to push the boats through overhanging branches, unperturbed by the caimans that occasionally plopped into its murky depths from steep, muddy banks. And they soon spotted signs of the isolated group we were looking for – plant stalks bent down at a 45-degree angle, called *quebradas*, or 'breaks', that these people had left to

Introduction: Into The Forest

mark their trail. Pereira examined them closely to confirm that the bending was deliberate and they hadn't been broken by a falling branch or a tapir that had left teeth marks.

'This is people,' he said. 'One month.'

Alcino Marubo – a serious, helpful man who wanted more involvement in protecting his reserve – took photos with Pereira's camera while he marked points on his GPS. They found more bent branches nearby, but these were a year old. Some had been snapped with a machete, others by hand. Nobody else hunted here, confirming that *isolados* used this route. 'This is strong information,' Pereira said. 'We were lucky.'

We hiked for days through dark, dense jungle. The ground was soft, damp, slick with sticky mud or knee-deep in brackish water. It was like trekking across a compost heap. I clutched at trees and branches to struggle up steep, slithery inclines, avoiding those with long needles or sharp, spiky bumps, nearly walking into a wasps' nest before Takvan alerted me. He stopped at one point and sniffed around him. A jaguar had recently been here, he said in his broken Portuguese. It was as if I was careering blindly through a one-dimensional, black and white forest, while he walked through a colourful universe cascading with information.

The team crossed muddy brown rivers, balancing on slippery logs that had fallen across them, using stabilising poles cut from the trees. I found these crossings nerve-wracking. After nearly falling from a couple of them, and half-falling backwards into a flowing river, I baulked at one dauntingly high log and decided to slide slowly across it on my backside. This delighted the Korubo boys, who roared with laughter as I painstakingly eased my butt over a knot on the log, some evil-looking, bright orange mushrooms and a black fungus as wide as my palm. From then on, I gulped down my pride and let someone carry my rucksack over these treacherous natural bridges which, frustratingly, Gary also stepped easily across.

Where I floundered, slipped and tumbled, the Indigenous people walked confidently. They were the ninja warriors of this forest. Manipulating razor-sharp machetes like a chef with a kitchen knife, they sliced trails through thick, tangled forest. They hunted, skinned, dismembered and roasted monkeys, sloth, tortoise, a wild turkey called *mutum*, and wild boar known as peccary. They fished piranhas from caiman-infested rivers and made camp in under an hour, clearing undergrowth and cutting down saplings to make the wooden frames that tarpaulins are stretched over and hammocks hung underneath, often in torrential rain.

And they knew all the forest's secrets. Where we saw mud and trees, they found nuts and fruit like cacao – the delicious yellow fruit with white, pulpy flesh whose seeds, when roasted, make chocolate. One morning, the two Korubo boys ran towards us, laughing and shouting, beating a beehive to chase away its inhabitants, before sharing its rusty-red honeycomb, dribbling with sweet, wild honey – a delicious glucose rush to beat the energy bars Gary and I had forgotten in a hotel room. Hundreds of kilometres from the nearest town, days of walking from the closest village, this for me was the wild, untamed Amazon forest of myth, TV nature shows and movies. For the Indigenous people of the Javari Valley, it was, quite simply, home.

For them, this forest was both a source of food and sustenance and an object of respect and reverence. Nature was not a postcard to be gazed it. It was a farm and a larder. It was intertwined with

Introduction: Into The Forest

their lives, like the creepers that wrapped around our feet as we stumbled through it. We were virtually living off the land, supplementing the rice, coffee and snacks the men were carrying with whatever they could hunt. Gary and I were utterly dependent on the Indigenous men and the prey they hunted each day, served up with rice, salt and a delicious chilli sauce someone had brought along. There was freshly killed *macaco barrigudo* – or 'fat-bellied monkey' – blackened on the fire to remove the fur then roasted, revealing its deliciously charred, fatty meat, like pancetta. The meat from barbecued peccaries tasted like roast pork. Piranhas were cooked on sticks. The Korubo ate all of it – meat, guts, brains – then collapsed into sleep straight after dinner.

Meals around the fire and a night's sleep or post-hiking rest in the hammock were moments of relief. At times, the forest seemed so thick and claustrophobic that, after days of hiking, it was a relief to see the river – and a horizon. It also provided moments of exquisite loveliness. When we came across a rare mahogany tree, towering majestically over a sun-dappled stretch of more spacious jungle, the men commemorated. The tree's rare presence meant good land, Pereira said.

'This is beautiful,' said Marcir Ferreira, a rare sentimental comment from this rugged frontiersman. A generous, resilient man, he insisted on sharing his talcum powder after every long day to ease sore feet.

This part of the Javari Valley had not always been dense, unpopulated forest. Before the area became an officially protected reserve in 2001, it was full of non-Indigenous *ribeirinhos* (riverside dwellers) who worked as loggers and rubber tappers for distant patrons. 'Many loggers came into this area, killed isolated Indigenous and took a lot of our riches,' said Aldeney da Silva, who lived in the Marubo village of Rio Novo that became our base during the expedition. He remembered life in the 1970s and the ongoing conflicts between tribal groups. 'There were wars between them, because of the non-Indigenous, rubber tappers and loggers.'

Marcir Ferreira moved to Javari when he was a child with his father, a 'rubber soldier' who had come from Ceará state on the other side of Brazil to tap rubber during the Second World War. 'It was very full,' Ferreira said. 'Many people lived by rubber tapping and logging.' When he was four, his family was attacked by Indigenous *isolados*, who set their house on fire. Ferreira was hit by an arrow that went through skin on the top of his head. His cousin, a young woman, was hit on the bottom, the shoulder and the hand. 'We hid behind a banana tree,' he said. Years later, he found out that his six-month-old niece had died after being shot by an arrow in the neck.

The Korubo have a different take on their warlike reputation. They were defending their land. 'We lived in the forest. There were no whites. When the whites came, we clubbed them,' explained a man I met in the Korubo village of Vuku Maë. Reports of contact with the Korubo go back to the 1920s, and 40 Korubo were reportedly killed by a Peruvian group accompanied by Ticuna Indigenous people in 1928. Throughout the 1970s, 1980s and 1990s, Korubo people were involved in conflict with non-Indigenous settlers, FUNAI staff and employees of the state oil company, Petrobras, which was prospecting for oil in the reserve. They are said to have killed around eleven whites. Nobody knows how many Korubo died but it is likely to be many

Introduction: Into The Forest

more. An untold number of Korubo were killed in an attack by Indigenous rubber tappers in 1979. Two years later, the leader of that group distributed poisoned farinha – the manioc flour that is a staple of Brazilian meals – to a Korubo community.

Stories like these are typical of the Amazon's bloody history and essential to understanding its present and its future. Gifts of poisoned food were just one of the techniques used to kill Indigenous communities. During the Amazon rubber boom of the late nineteenth and early twentieth centuries, the London-registered Peruvian Amazon Company was denounced for the enslavement, torture and murder of tens of thousands of Peruvian Indigenous people on its rubber plantations. In the 1940s, one Brazilian rubber company reportedly punished Indigenous workers who failed to meet production targets by first cutting off one ear, then the second ear, then with execution.

In an exposé written by Norman Lewis, published in the UK's *Sunday Times* magazine in 1969 and entitled simply 'Genocide', a federal police detective called Neves da Costa Vale told the journalist that 'hundreds of Indians were being enslaved on both sides of the border' with Peru. The feature centred on an extensive report by Brazilian prosecutor Jader de Figueiredo into 134 employees of the government's Indian Protection Service. Indigenous communities had been wiped out after being given clothing impregnated with smallpox and poisoned food. Some were gunned down by machine gun fire. Women were stolen as wives. Tortures were horrific and commonplace. 'The tribes had all been virtually exterminated, not *despite* all the efforts of the Indian Protection Service, but with its *connivance* – often its ardent cooperation,' Lewis wrote (his italics). The service was dissolved and replaced by FUNAI.

The Brazilian military has never come clean about its atrocities while in power – and a 1979 Amnesty Law protected anyone, on the left or right, who committed crimes during its rule from prosecution. The Figueiredo Report then mysteriously disappeared, only to resurface in 2013, after leftist president Dilma Rousseff set up a National Truth Commission to investigate crimes and human abuses under Brazil's military dictatorship and previous governments. She herself had been a member of an armed Marxist guerrilla group that opposed the dictatorship, and was brutally tortured in prison.

'The Indigenous peoples suffered grave violations of their human rights between 1946 and 1988,' the commission concluded. 'These violations were neither sporadic nor accidental: they were systemic, as they resulted directly from structural State policies.' It concluded that at least 8,350 Indigenous people had died, either as a result of direct action by government agents or their omission. Direct actions – murders and massacres – were concentrated during the military dictatorship, from 1964 to 1985; omissions were more common before it. In just one of the atrocities listed, 33 Indigenous people were killed when a helicopter poured a deadly powder on them. Their village was near a highway the army was building. Just one person survived.

In the run-up to Bolsonaro's election in 2019, a year after my trip to the Javari Valley, invasions of forest reserves, deforestation and fires which had already been rising during Rousseff's presidency exploded. Loggers, *garimpeiros* – as the gold miners are called – and land grabbers felt emboldened by Bolsonaro's rhetoric. I know this because they told me – including on one reporting trip to the

Introduction: Into The Forest

Amazon state of Rondônia between election rounds in 2018, and another to wildcat mining sites in deep jungle in the Yanomami Indigenous Land in 2019. These were men from rough, frontier towns that offered few jobs beyond low-paid, manual labour. They had little awareness or understanding of global environmental issues – an ignorance bolstered by the fake news and propaganda that Bolsonaristas bombarded social media with. They wanted to work and faced repression from environmental authorities, the police or the army. They believed that Bolsonaro was on their side and had voted for him because, they explained, he was going to legalise their work.

After the election of Bolsonaro, the Javari Valley saw even more fishing gangs and began to come under pressure from Evangelical Christian missionaries – conservative Evangelical churches are another of Bolsonaro's power bases, along with the army, police and the powerful agribusiness lobby, known as the *ruralistas*. And the *ruralistas* were wielding even more influence over FUNAI.

In 2018, Bruno Pereira had been put in charge of the agency's department for isolated and recently contacted Indigenous, a prestigious and highly sensitive role. Some time later, Marcelo Xavier da Silva, a federal police officer, took over as FUNAI president. Its employees were appalled because da Silva had worked on a controversial Congress inquiry helmed by a leading *ruralista* lawmaker in 2016 which called for FUNAI and NGO employees to be prosecuted. FUNAI was an 'operational arm of external interests' and 'an amalgam of private interests and ideological objectives', it said, calling for commercial farming to be allowed on Indigenous territories where it is currently prohibited. I wrote about the report at the time. Back then, it sounded bizarre and extremist to me, its convoluted argument that US farmers wanted the Amazon conserved to protect their own farming interests a dangerous conspiracy. But such seemingly outlandish, illogical thinking was soon to become the new normal in Brazil. The far right had taken over, and anyone disagreeing with its more extreme believers was dubbed a 'communist'.

With Xavier da Silva in charge, Pereira was abruptly demoted (as a federal employee, he couldn't be sacked very easily). No explanations were given. Indigenous experts warned that Brazil's isolated peoples faced 'genocide' in an open letter.

If the seeds for this book were planted on my trip to Javari in 2018, they took firm root after Bolsonaro's win, when I visited the dusty settlers' town of Novo Progresso on the other side of the Amazon in the state of Pará – a town where Jair Bolsonaro got 78 per cent of the vote.

Novo Progresso sits beside the BR-163, a highway linking the soya- and grain-producing heartlands of Brazil's central-west to Amazon ports like Santarém and Mirituba. At its entrance, a sign read 'Town of Development', near a huge Bolsonaro billboard with the slogan: 'On the development route'. But a significant, if unquantifiable, amount of the 'development' that drives this busy community of 26,000 people is illegal. There are dozens of secretive gold shops in its streets and the *garimpeiros* that supply them mostly ply their pollutant-heavy trade in protected and Indigenous territory where it is prohibited. Novo Progresso's herd of over half a million cattle has risen about 50 per cent over the last decade. Yet the town is in the midst of protected forests and Indigenous territories where commercial farming, on paper at least, is prohibited.

Introduction: Into The Forest

Amazon farmers and land grabbers traditionally burn cleared and deforested land in the dry season, but in August 2019, the number of fires roared to a nine-year high and provoked an international crisis for Bolsonaro's government. He tried to blame NGOs and then actor and environmentalist Leonardo DiCaprio, without producing any evidence. Police and prosecutors launched an investigation into what was dubbed 'Fire Day' – when farmers in the Novo Progresso area allegedly coordinated setting fires on 10–11 August, two days in which fires in the area tripled. It was widely interpreted as an act of defiance against environmental regulations and a show of support for Bolsonaro's stance that the Amazon was there to be exploited. Behind this was an argument on the right that rural livelihoods were being held back by unfair protections of the forest, which were put in place to pander to conservation groups and foreign governments 'We need to show the president that we want to work,' one of the farmers involved told a local news site, which revealed the coordinated action. The site's owner and reporter, Adecio Piran, later received death threats for revealing 'Fire Day'.

Novo Progresso is located next to the Jamanxim National Forest – a 1.3 million hectare, federally protected reserve that is one of the most deforested in Brazil. It was created in 2006 to slow rampant deforestation in the area and is run by the Chico Mendes Institute (ICMBio), a federal government environment agency. Officially, farming is not allowed inside – but locals have never accepted the reserve's existence. 'Farmers see the Jamanxim National Forest as a deterrent to the development of cattle raising in the area,' a 2009 ICMBio study said, concluding that two-thirds of those inside the forest entered shortly before or after the reserve was created and that most farmers lived elsewhere – many in different states. The state of Pará has Brazil's second-biggest cattle herd, 24 million cows, up 33 per cent in ten years. Since 2006, Pará has also been the Amazon state that consistently deforested most land, according to government satellite data. Cattle ranching is the biggest driver of deforestation across the Amazon. Fires are used to clear land once the valuable trees have been removed and sold. Cattle are put on that land to consolidate possession.

A few months after 'Fire Day', I went with Daniel Camargos, a reporter with Brazilian agency *Réporter Brasil*, and photographer João Laet to tour on bumpy dirt roads the Jamanxim Forest and a nearby reserve, the Springs of Serra do Cachimbo, on assignment for *The Guardian* and *Réporter Brasil*. We saw how swathes of what was once protected forest had become pasture, where cows grazed amid the blackened shells of charred tree trunks. Wildlife survived as best it could.

Driving down a dirt road in the Cachimbo reserve one morning, we passed a man standing lookout beside a motorbike while chainsaws squealed in the trees. It was a tense moment: running into loggers in reserves, who are often armed, can be dangerous. Just inside the Jamanxim Forest, recently burnt forest on one farm was still smouldering the morning when we arrived. And so it went on.

A farm we visited in the Serra do Cachimbo reserve had a long list of fines for environmental offences. These penalties are one of the government's main tools to limit deforestation, though their effectiveness is mixed. Many ranchers refuse to pay and rack up huge unpaid fines, with seemingly little direct consequences. But these records serve as an official black mark against the landowners,

Introduction: Into The Forest

which can – if supported by other official agencies and financial institutions – be used to limit their access to loans and markets. That, in turn, depends on the government in power and the perspicacity of traders, retailers and their regulators.

The farm in the reserve was owned by Paulo Parazzi, an official from the environment agency of the government of Paraná state, thousands of kilometres away in the south of Brazil. He refused to talk to me when I called him at work. Another 6,000-hectare farm, where cattle grazed among charred logs in fields surrounded by a ring of forest, was registered to André Ferri, who also lived far away in Paraná state. He owed millions of pounds in fines, but the Brazilian justice system had never been able to find him. My colleague Daniel Camargos had no problem finding his number and made contact, but he didn't want to talk to us either.

Ruralistas sometimes argue that fires and deforestation are done by small farmers struggling to make a living. However, several studies, including one called 'Rainforest Mafias' by Human Rights Watch, have shown that often, such destruction is driven by criminal gangs. Protected forests like Jamanxim and Serra do Cachimbo are subject to intense real estate speculation from investors in other states, betting that in future, their reserve status will be revoked. People with land inside the reserves invariably say they were there before the reserve was set up, which – often dubiously – enables them to claim they have a prior claim to the land, and that they are victims of the state rather than perpetrators of crime. Environment agencies fine them – but, again, nobody pays. One large patch of recently burnt forest we visited in the Serra do Cachimbo reserve was registered to Nair Petry, a woman who said she had been there since 2001. Not long before it was set on fire, she had offered the land for sale on her Facebook page for around £500,000 – though she later denied this in a phone interview with Camargos. As she herself confirmed, the new owners of land like this inside reserves usually put cattle on it quickly. Those who put cattle on newly deforested land, or farmers who have extended their own land by illegally felling trees, can find ways to sell their cattle.

In 2009, Greenpeace produced a blistering report called 'Slaughtering the Amazon', which exposed how much deforestation the cattle industry was involved in because it was buying from farmers illegally felling trees. Later that year, Brazilian meat companies inked deals with Greenpeace and, later, federal prosecutors, in which they committed to not buying cattle from farms that had been fined or had areas embargoed – that is, where production on the land had been prohibited – because of illegal deforestation. These companies have since spent heavily on sophisticated monitoring systems to stop that happening, as Brazilian-owned JBS, the world's biggest meat company, explained in an email to me.

'The JBS monitoring system in the Amazon covers more than 280,000 square miles, an area larger than Germany, and assesses more than 50,000 potential cattle-supplying farms every day,' the company said. 'To date, we have blocked more than 8,000 cattle-supplying farms due to noncompliance.'

But there is a way around it. The process is sometimes called 'cattle laundering' and it's helped by the fact that in Brazil, few farms handle the whole life cycle of a cow, from birth to delivery to slaughterhouse. Instead, there are farms that breed and birth cattle, others that fatten them and others that 'finish' them, ready to supply slaughterhouses.

Introduction: Into The Forest

The monitoring systems used by companies like JBS, and its smaller rivals Marfrig and Minerva, only check the direct suppliers – those at the end of the chain selling direct to slaughterhouses – and not those that supplied them, the so-called indirect suppliers.

In their deals with Greenpeace and Federal Prosecutors, these leading meat companies promised to have full control over their indirect suppliers by 2011. More than a decade later, they have still to fulfil that promise. The reality they operate under remains as murky as Amazon mud.

Residents of Novo Progresso we spoke to explained how they saw themselves as hard-working pioneers taming an Amazon wild west. 'Wood, gold prospecting and now cattle made this town,' said Jadir Rosa, a quiet, serious mechanic who had moved from Paraná state in the south of Brazil and was lunching in the town's market one day. Rosa supported Bolsonaro and shared his government's scepticism over climate science. 'Global warming does not exist,' he said. He said people in the town got a lot of information from the messaging app WhatsApp.

Novo Progresso's unofficial town spokesman is Agamenon Menezes, an influential and combative former president of the local rural producers' union. He was interviewed by police during the 'Fire Day' investigation and his computer seized, but he denied any involvement. Dry and dismissive of journalists, Menezes made a vague threat under his breath as we talked and raged quietly against a US reporter he had spoken to because he did not like the way he had been described. He said 'Fire Day' was a media invention to attack Bolsonaro. He also said that 35,000 'serious' Brazilian scientists had disproved climate change science (I was unable to find any evidence of them, despite the large number). More interestingly, he explained that Bolsonaro's popularity in Novo Progresso was because he challenged environmental officials and attacked regulations that Menezes said stopped honest Amazon people working.

'They have to eat, they have to produce food. So they work illegally,' he said. 'Nobody wants to be illegal as well. They want to work legally.' Fires are used to clear land for pasture, which is later used for agriculture, he explained. 'You get an area of dense forest and deforest it,' he said. 'You need to burn this wood.'

As 'Fire Day' showed, the argument that Amazon people want to work honestly and that environmentalists and NGOs are part of a global conspiracy stopping them from doing so is widely used in the Amazon. The Bolsonaro government itself makes these arguments and key ministers dispute that global warming has a human cause. Combating them effectively is crucial to saving the Amazon because if cutting it down makes no difference to anything, as Menezes argues, why save it? All you are doing is stopping good, honest folk from working.

The problem is that the work of Novo Progresso's townsfolk has an impact on all of our lives – and on all of our futures. That is because the fires and deforestation in places like this are taking the Amazon closer to what leading climate scientists like Carlos Nobre and Thomas Lovejoy have called its 'tipping point', beyond which it no longer produces enough moisture to maintain itself and 'dies back' to become a semi-arid savannah. In December 2019, they wrote in *Science* magazine: 'The precious Amazon is teetering on the edge of functional destruction and, with it, so are we. The

Introduction: Into The Forest

moisture of the Amazon is not confined to the basin but is a core and integral part of the continental climate system with specific benefits for critical Brazilian agriculture in the south.' Seventeen per cent of the entire Amazon basin and nearly 20 per cent of the Brazilian Amazon is already deforested, bringing the tipping point to within 15–20 years, or even less. 'The tipping point is here, it is now. The peoples and leaders of the Amazon countries together have the power, the science, and the tools to avoid a continental-scale, indeed, a global environmental disaster. Together, we need the will and imagination to tip the direction of change in favour of a sustainable Amazon,' wrote Nobre and Lovejoy.

Two weeks after Agamenon Menezes told us that global warming did not exist, Carlos Nobre was in Princeton University, New Jersey, at an unusual conference called 'Amazonian Leapfrogging'. Increasingly, scientists like Nobre are not just analysing what is going wrong in the Amazon but researching ideas that could help it go right – such as by providing jobs and opportunities for people like those in Novo Progresso that do not involve destroying the forest. Princeton scientists are part of this movement. They have teamed with leading environmentalists and scientists from Rio de Janeiro's Pontifical Catholic University and other research institutes on a project called Amazon 2030, which analyses Amazon economy and business to suggest how it can be improved while leaving the forest standing. At the Princeton conference, climate change scientists presented the results of climate modelling to simulate a world in which the Amazon was completely or 50 per cent replaced by cattle pasture – not an unreasonable scenario, given that cattle ranching is blamed for 60–80 per cent of deforested Amazon land. As climate-change expert Stephen Pacala told

Brazilian magazine *piauí*, 'The increases in temperature will be catastrophic for Brazil.' It found that the Amazon could get up to 5C hotter. And it would rain 25 per cent less in Brazil, with dire consequences for agriculture and the wellbeing of humans and other species.

Unfortunately, neither the most well-respected scientists nor their painstakingly in-depth studies are of any interest to the far right in Brazil. Former environment minister Ricardo Salles called the debate over global warming a 'secondary issue'. Former foreign minister Ernesto Araújo described what he called *'climatismo'* – which you could roughly translate as 'climatism' – as part of a globalist, Marxist plot. 'Climatism is basically a globalist tactic to install fear to obtain more power,' he wrote. During press conferences in Rio de Janeiro in 2019, I asked two other leading figures in the Bolsonaro-era government if they believed that global warming had a human cause. Neither was convinced. Former vice-president General Hamilton Mourão, who headed Brazil's 'Amazon Council', said there was still a debate as to whether climate change was a 'seasonal change'. Former finance minister Paulo Guedes told me he thought 'there is still a precarious scientific basis' for climate change science.

The UN's Intergovernmental Panel on Climate Change, which involves 1,300 scientists from all over the world, believes there is a greater than 95 per cent chance that human behaviour since 1950 has warmed the planet. NASA agrees. So did an exhaustive report from 13 US government agencies. Though former US president Donald Trump disagreed – and Bolsonaro was closely allied to him. This put Brazil under a lot of pressure to take action on its main source of climate-destabilising emissions: deforestation.

Introduction: Into The Forest

As awareness of the climate emergency increases, together with extreme events like floods and fires, the clamour for change is coming from investors, businesses and customers. Big money has realised it's bad for business to be associated with environmentally destructive products and are lining up – on the surface at least – with environmentalists.

This book doesn't just describe the destruction of the Amazon; it looks for ways to stop it and to heal it. During my research, I have met Indigenous people, other forest communities, social activists, entrepreneurs, environmentalists, scientists, economists, anthropologists and farmers, all of whom know and understand the Amazon intimately and have innovative solutions for the millions of people who live there, whether they are involved in destructive activities like logging, illegal gold mining and cattle farming or suffering their impacts.

I also want to show readers what they can do: how the rising pressure that investors are putting on Brazil is making a difference, as is consumer anger over the way deforestation is caused by meat multinationals' willingness to buy from Amazon cattle farmers who have committed environmental crimes. Consumers and companies can force change when they vote with their wallets. As this book will explain, sustainability is no longer just a moral obligation to future generations – it's now part of the business bottom line.

Saving the Amazon requires looking at the rainforest as an asset, not an obstacle to progress, as many Brazilians have historically done. It involves developing areas like biotechnology and sustainable land management, and recruiting local communities into environmental protection using international funding. It involves bold international initiatives and grassroots successes. It means more Indigenous and protected reserves. With so many possibilities, the book will focus on ten areas where exciting work is already being done.

The people getting rich by exploiting and destroying the Amazon have one thing in common: very few of them come from its Indigenous communities. Of course, there are Indigenous people involved in logging, wildcat mining and other illegal activities, but the powerful and shadowy figures behind these operations are almost always non-Indigenous. Throughout this book, I will keep returning to the Amazon's Indigenous people, its original inhabitants, and the lessons they have for saving it. And I will keep returning to the words of their leaders and thinkers, who have much to tell us. One of the most important is Davi Kopenawa, a shaman and leader from the Yanomami people, whose reserve has been overrun by mostly non-Indigenous wildcat gold miners for decades – who have increased under Bolsonaro's presidency, as I found on a visit in 2019. 'Indigenous people will never get rich. It's the whites, the big businessmen who get rich,' he told me in 2018. Maneose Yanomama, 54, a Yanomami shaman in the Sikamapiu community told me that the spirits of nature were sounding the alarm: 'The whites are getting closer. They are damaging our land, they are destroying our rivers, they are ruining our forests. Nature is very scared.'

People need to learn from Indigenous peoples that only collective, community thinking, not individual greed, can save the Amazon. We need to pull together, not pull apart. And the world needs to understand how the riches, knowledge and biological

Introduction: Into The Forest

super-computing power in the forest makes it infinitely more valuable long term than if turned into dry, unproductive cattle pasture. Which brings us back to the jararaca snake that Takvan Korubo killed in the Javari Valley.

In the 1960s, while based in a university in Ribeirão Preto in southeast Brazil, scientist Sergio Henrique Ferreira was able to isolate a peptide from the venom of the jararaca snake. He later worked in the UK where his colleague John Vane and other scientists used that peptide to develop the active ingredient in captopril, the first angiotensin-converting enzyme (ACE) inhibitor, whose effects on blood pressure mimicked those of the snake's venom. This led to the development of drugs to treat high blood pressure and heart failure. Vane later won a Nobel Prize for his work – and Ferreira was at the ceremony. The discovery is regarded as one of the most important breakthroughs in the treatment of heart problems and Ferreira was inducted into the prestigious Brazilian Academy of Science in 1984. He died in 2016.

How many more lifesaving peptides might there be in the Amazon? Nobody knows because most of the region's species are still undiscovered. It's one of the myriad reasons the hostile habitat of the jararaca and other species should be preserved. And that, essentially, is what this book is about: not just how to save the Amazon, but also why.

Brazil

CHAPTER ONE

Laying Down the Law: Political leadership

Dom Phillips
Mamirauá, Amazonas State

'Friends, if you bring down two or three helicopters, whoever they belong to – the government or the army – [it] will have to stop.'

On a hot, dry October morning, two small wooden boats chugged up the River Japurá in the protected Mamirauá Reserve in the Brazilian Amazon, each pushing a much larger, heavily laden barge before it. One of the vessels was a long, black cargo carrier, laden with metal tubes and plastic tanks of fuel; the other bore a two-storey green shed with a water tank on its roof, clothes drying on a second-floor veranda, and conveyor belts, suction tubes and yellow machinery stacked below. Two men chatted near its prow as we sped past in a small aluminium speedboat. This was a gold mining barge, heading upriver towards the municipality of Japurá, situated less than 300km from the Colombian border in a region known for its wildcat gold mining – called *garimpo* – and the violence and crime that invariably comes with it. This ramshackle river traffic was completely illegal. And yet it was far from an uncommon sight.

You don't have to look far in the Brazilian Amazon to find people breaking the law. They're all over the place. The articulated truck, loaded with enormous logs and rumbling down an Amazon highway at night. The sawmill operating just outside a protected Indigenous reserve, far from any legally logged trees. The nurse at an isolated health post on an Indigenous reserve, obliged to tend bullet wounds and snake bites in miners who shouldn't even be there. The father and son casually building a two-storey house inside a protected reserve where it is forbidden to do so. The sprawling cattle ranch whose skinny animals range among charred tree trunks, in a clearing inside a forest where commercial farming is prohibited, with tens of millions of dollars in fines that will never be paid. The wrecked flatbed trucks, lacking doors and bonnets and licence plates, like props from the *Mad Max* films, engines howling as they ferry tree trunks down forest tracks. They will later be left outside garages in busy towns in plain view of police who will never bother them.

These are all sights I have seen at different moments in the Brazilian Amazon. At times, it seems like the law doesn't really exist or exists in a separate dimension, where people only need to pay it passing attention from time to time. Rich and powerful farmers find ways to launder their illegally raised cattle so they can sell to multinational meat companies who in turn export to foreign supermarkets with imposing sustainability commitments. Land grabbers pay hitmen to murder poor farmers standing between them and forests they want to steal. The leaders of logging gangs spend a few months in prison then go back to their businesses – applauded or even elected to public office by their local communities. Precious hardwood may be legally logged, but then the permits for this are used to launder illegal wood, cut from protected reserves and Indigenous land, another natural heritage which the Brazilian people is steadily stealing from itself while the state stands by, scratching its armpits and yawning.

It's just as easy to see illegal mining barges lining rivers. Because even though the *garimpeiros* clear forests, suck gold out of river mud, dump the deadly mercury they use to separate it out into rivers, where it gets into food chains, while bringing violence, prostitution and drugs to remote communities, they can be seen all over the Amazon. Even when federal police or the army bust the mines, they generally let the *garimpeiros* go. It's difficult to think of anywhere else outside of a war zone where it's so easy to find people blatantly breaking the law.

While these illegal activities have always been present in the Amazon and, until a few decades ago, were not just legal in many cases but actively encouraged by the government and banks, they have worsened over recent years. And they began accelerating out of control when Brazil's hard right president Jair Bolsonaro took office in January 2019. As far as he and his followers – many of whom are loggers, *garimpeiros* and land grabbers – are concerned, there is nothing at all to be worried about. Meanwhile, the evidence accumulates that some of the ecosystems that make up the Amazon have become severely impacted. Parts of the rainforest now emit more carbon than they absorb. Rainy seasons have got shorter in 'transition zones' on the edge of the forest. Specific species of trees and plants have begun to disappear. Others have become more dominant.

So what can be done? Ask people in the Amazon who are worried about saving it and they tell you the same thing: there is no silver bullet. No magic solution. Instead, there are a range of things that would help, from government policy initiatives to foreign financial assistance. And many of the Amazon experts I talked to for this book said the same thing.

To start with, they said, it is necessary for the state to take the lead. This is what some governments have tried to do, following a concerted, organised effort that coordinated dozens of government ministries together with environment agencies, federal police, the army and even the Central Bank. This was hugely effective in Lula's first two administrations, from 2003 to 2010, when Brazil got deforestation under control. In his third term, the country is trying to repeat the successful strategies employed back then. It starts with the state asserting itself and cracking down on illegal activities. This approach is generally referred to as 'command and control'.

'We were able to get great results with the increase in investments in environmental control. We started to get more structure, we started to have aircraft which we hadn't had, we invested in vehicles,' Wallace Lopes, an inspector who joined federal environment agency Ibama in 2009, told me. 'We were carrying out a plan.'

Lopes works extensively on Amazon operations. He has a dangerous job. Ibama staff like him are legally allowed to carry firearms and are accompanied by armed police on operations – sometimes swooping down in helicopters to raid a *garimpeiro* or logging camp. Their vehicles and bases have been set on fire by angry mobs in Amazon towns where locals regard them as a threat to their livelihoods. And yet Lopes was anything but authoritarian. Wearing a Gibson guitar T-shirt with the slogan 'choose your weapons', the bearded environment enforcer came over as a genial and thoughtful character when he spoke to me on a video call from Boa Vista, capital of the Amazon state of Roraima, where he was on a mission combating *garimpo*.

Wildcat gold mining is so entrenched into Roraima's economy that there is even a road in Boa Vista known as 'Gold Street', lined with nondescript, shabby jewellery shops that will weigh and buy a fistful of raw gold with no questions asked. Shopkeepers told me as much once when I walked down it with a photographer, but clammed up when we asked for an interview. Local politicians have long been suspected of involvement. According to a 2021 police investigation, Jalser Renier, the former president of the state's legislature, ran a paramilitary gang with corrupt police officers that investigators believed kidnapped and tortured a critical journalist and supplied *garimpeiros* with weapons. Renier, who was suspended from the legislature's presidency, has denied the allegations. In January 2021, under his presidency, the legislature passed a law facilitating *garimpo*, and even the use of mercury, although Brazil's Supreme Court overturned the measure.

Lopes and his Ibama colleagues were staying in a safe house, not a hotel. They avoided wearing any Ibama insignia or uniforms and did not go out at night. He had blurred the background during our call and other colleagues who occasionally made comments stayed out of his laptop's camera view. Their mission was focused on cutting off the *garimpeiros*' air and food supply networks. They had blown up *garimpeiro* helicopters. Lopes sent me audio recordings that colleagues had harvested from *garimpeiro* WhatsApp groups that came thick with threats. The *garimpeiros* were enraged and wanted to shoot down Ibama helicopters and kill anybody destroying their equipment. Their argument was that they were working honestly to feed their families.

'These vagabonds arrive, set things on fire and nothing happens to them. Friends, if you bring down two or three helicopters, whoever they belong to – the government or the army – [it] will have to stop. It will go on national media,' a man said in one voicenote. 'Gold comes from God's nature. And if we need it, where do we have to go? Where it is. And nobody is stealing, they are working.' He continued the theme on another recording: 'Everybody has to unite, and when the helicopter comes wanting to burn things, strangle it with bullets.'

Under Bolsonaro's government, Ibama staff were banned from talking to the media. Lopes was only able to give me an interview because he is a director of the federal environment workers' union, ASCEMA, and was speaking in a personal capacity. He recalled that in 2004, 27,772 square kilometres of Brazil's Amazon rainforest was destroyed, and how by 2012, that had fallen to 4,571 square kilometres. 'If we had the structure and the necessary staff, we could bring deforestation below 5,000 square kilometres in two years,' Lopes said. That almost proved to be the case in just one year after Lula returned to power.

At about 2am, on a balmy August night in 2014, the fire engine carrying the coffin arrived. It moved slowly through the crowds outside the ornate governor's palace in Recife, a rough, culturally vibrant city that is capital of the northeastern state of Pernambuco. Grief drenched the dry air. Grief for a popular politician who'd had, many here believed, a promising future. Eduardo Campos, the blue-eyed, patrician scion of a powerful political clan, had governed the state for eight years and had set his sights on the presidency. He had been polling at around 9 per cent and was in third place with less than two months to go before the election. He was seen as the 'third way' candidate, seeking to present himself as a reasonable voice and effective manager who could reconcile growing the economy and helping poorer Brazilians while protecting the environment.

But Campos and six others had died in a small plane crash the previous day and now his sons punched the air from the fire engine roof as crowds chanted his name. When his vice-presidential candidate, Marina Silva – a softly spoken environmentalist from the Amazon state of Acre – appeared, they chanted her name too, but less forcibly. They cheered again when she appeared at the requiem mass, shown on giant screens to a multitude which had swelled to an estimated 130,000, many of whom queued for hours to see Campos's coffin. Marina Silva was now expected to take over from Campos as the main candidate, and her name kept coming up as I circled the crowd, interviewing mourners for a story. Could she carry the vote?

Silva had made her name as an environmental activist, served as a senator and minister, and came third in the elections four years previously, in 2010, getting 20 million votes before being beaten by Workers' Party candidate Dilma Rousseff, groomed as the successor to the then wildly popular, two-time Workers' Party president Luiz Inácio Lula da Silva. Both women had served as ministers in Lula's government.

Four years later, she returned as Campos's vice-presidential candidate. She was back in the ring and would face Rousseff again. The women knew each other well. Each saw the environment very differently and had clashed in government over their different visions. Dilma Rousseff believed in state-driven development.

Marina Silva told me: 'I became a defender of the Amazon, of the environment, of the forest, practically as a child.'

It is neither a coincidence nor a surprise that the Brazilian politician who has arguably done more in recent history than any other to protect the Amazon rainforest was born and raised there. A self-contained, sombre and composed woman who is clearly hugely passionate about her work, Silva is very different from the pugilistic, high-octane personalities who dominate so much of Brazilian politics – most of whom are men. She is an Evangelical Christian like so many in the Amazon, but unlike many of her religion, she is also a fierce critic of the Bolsonarist farming, logging and mining lobbies that threaten the rainforest's survival.

I first interviewed her in 2014, after Eduardo Campos's death had rocketed her back into the race at a point where polls had her winning the run-off vote against the incumbent Dilma Rousseff. Marina Silva had been polling second early in the campaign, but she was foiled by bureaucratic hurdles from launching her own Sustainability Network party in time.

Wearing a cream linen trouser suit with her hair in the trademark bun, she spoke slowly and clearly as she told me how when she was growing up in a wooden house on stilts in the rainforest, in an isolated, riverside community in the forest in Acre, her father used to listen to the BBC World Service in Portuguese and Voice of America on a radio which he'd put somewhere high out of the reach of her and her siblings. Her parents had eleven children in all, but three died young.

'My father was addicted to news,' she said. 'I learnt a lot, yes, from the radio, too. To have an idea that a world existed beyond the place we lived in.'

Laying Down the Law: Political Leadership

In 2008, Silva went to London to receive the World Wildlife Fund's Duke of Edinburgh medal for her work protecting the Amazon forest as environment minister. She took photos outside Broadcasting House headquarters and was interviewed by the BBC. 'It was like I had gone inside the radio, the BBC. It was such a strange thing,' she said.

I interviewed Marina Silva again for this book. She was 63 by then and one of a group of former environment ministers issuing coordinated critiques of the Bolsonaro government's onslaught on Brazil's environment. With 2.1 million followers on Twitter (subsequently X) and another 270,000 on Instagram, it seemed to me that Silva had returned to being an activist, rather than a politician. Perhaps she would argue she always was. 'My passion for the forest, for the Amazon, for its traditional peoples has always been with me,' she said.

During the Second World War, some 55,000 migrants were brought from Brazil's poor, semi-arid northeast to work as rubber tappers in the Amazon to supply rubber for the USA's war effort. Their numbers were decimated by diseases like malaria and yellow fever and around half died, but their descendants still live in Amazon cities and communities. When he left his native state of Ceará, Silva's father, Pedro, was just 17, too young to be a 'rubber soldier', so he worked in the kitchen of a boat carrying these so-called soldiers. He became a rubber tapper and stayed. He met Marina's mother, Maria, who was also from Ceará, in Acre. Like many rubber tappers, Pedro Silva was obliged to sell all the rubber he produced to one boss, and to buy supplies sold at inflated prices by the same boss. 'This created a system of semi-slavery,' Silva said.

From around the age of five, Silva lived with her grandmother – a traditional midwife, an unmarried aunt, an uncle who was a shaman and other older relatives, 15 minutes' walk from her parents' house. Her mother died when she was 15. As a child, she and her six sisters and one brother got up at 4am to cut rubber trees, then collected the latex in the afternoon. Although she couldn't read, she described an education rich in natural remedies and forest folklore. It was a childhood of scarcity and abundance, she told me. There was no school, no health post, and if someone got seriously sick they were in trouble. But the family had the chickens and pigs they raised, the fish they caught, the melons, rice, manioc and pumpkins they grew. 'We had subsistence agriculture that did not impact the forest,' she said. 'We didn't go hungry, but we had problems getting medicine and clothes were very expensive.'

Marina caught malaria several times and leishmaniasis. Aged 16, she went to the state capital of Rio Branco to get treatment for hepatitis. There, she learnt to read and began studying to become a nun. She heard about 'liberation theology', a Latin American, Catholic movement, which preached freedom from social, political and economic oppression, from bishop Dom Moacyr Grechi, who held mass at her convent. Inspired, she joined Catholic social groups, called 'basic ecclesial communities', that had grown up in countries like Brazil, then living under a right-wing military dictatorship. At 17, she met Chico Mendes, a rubber tapper, union leader and environmentalist, who was fighting against the destruction of the forests rubber tappers like him depended on for their living, pitting them against ranchers set on taking over their land and razing the forest for cattle, either by buying them out or forcibly expelling them. Marina Silva told me she had realised that she was already an environmentalist, through birth and forest education.

'Together with my parents and brothers and sisters, I took my first survival, my first entertainment from the forest, where I learnt the first mysteries of nature. I've always carried this with me,' she said. 'I discovered that what we were learning in theory we already did in practice.'

She joined Mendes on passive resistance demonstrations called *embates* – or 'collisions' – where rubber tappers and activists formed human chains to stop the trees being destroyed. But his high-profile opposition to farmers and Amazon destruction made Mendes a target and in 1988, while under police protection, he was killed by a shotgun blast. He was just 44. In 1990, rancher Darly Alves da Silva and his son Darci Alves were convicted of the murder and imprisoned – Darly for ordering it and Darci for firing the shot.

In the ten years following her eighteenth birthday, six people Marina Silva knew were murdered, including Mendes, all of whom were involved in environmental or related work. Defending the Amazon forest is just as dangerous today as it was then: according to the non-profit group Global Witness, in 2022 alone, 24 Brazilian environment and land defenders were murdered.

Silva went into politics and served as a councillor in Acre's state capital Rio Branco, before becoming a state deputy and then a senator for Acre. She had just been re-elected with three times her initial vote when Brazil's new president Lula announced her as environment minister. She only found out when Congress reporters rushed to her for her reaction.

A former metalworker and union leader who had led a series of strikes in the dying years of Brazil's military dictatorship, Lula founded the Workers' Party in 1980 and was elected president in 2002 on his fourth attempt. Like Marina Silva, he had also grown up in poverty, one of eight children who emigrated with his family from the dusty backlands of Pernambuco in the northeast to São Paulo. In 2002, after losing the presidential election three times in a row, he softened his message, wrote an open letter to calm financial markets worried about what he might do if he won, and made a textile tycoon turned politician his vice-president. He was elected on a wave of optimism and hope for change in a deeply unequal country – a country that in many ways, during his first two four-year terms, he changed beyond recognition.

His government's cash transfer policy for poorer Brazilians, called *Bolsa Família*, or Family Purse, helped 30 million people escape poverty. Quotas for Black and Indigenous students enabled people to enter universities and jobs that had been closed to them before. This was a new government and it wanted to make its mark. Silva said she would only become environment minister if she could choose her own team, many of whom were environmentalists she brought in from civil society; others she left in place because she considered them good at their jobs. 'It was an unprecedented opportunity,' she said.

Tasso Azevedo was one of the environmentalists who joined the ministry. He had trained as a forest engineer – to manage the resources of the forest – and founded the non-profit group Imaflora, which certifies sustainable products and works on sustainable forest management. We talked on a video call. Wearing his straight, neat hair in a side parting, a shirt and glasses, Tasso spent most of our conversation on foot, as if to keep in step with the flow of information that came spilling out of him. Joining the Environment Ministry was 'neither a surprise nor something planned, it was something that just happened,' he said. 'And I stayed.' Later he would become coordinator of the Mapbiomas project, which maps and monitors deforestation.

João Paulo Capobianco was a biologist who had helped found and run two of Brazil's most respected environmental NGOs – SOS Mata Atlântica (SOS Atlantic Forest) and Instituto Socioambiental (Socio-environmental Institute, ISA). At ISA, he had taken part in meetings with Lula before the election, and when his wife, an advertising executive, took a job in the capital, Brasília, he accepted the role of Marina Silva's deputy. 'There was euphoria,' he told me during another video call. 'There was a lot of expectation that it would be possible to put the main ideas of the socio-environmental movement into action.'

Capobianco was a thoughtful, serious man with a grey goatee who later wrote a doctorate on what the plan had achieved and why and what it had not. But he also had a sense of fun. Where Marina Silva was formal and discreet in her descriptions of political colleagues, Capobianco interspersed his tales of high-powered meetings with convincing impressions of the gruff, passionate, tough-talking Lula. He and Marina Silva both knew what their biggest challenge was. 'There was an issue that was, let's say, our Armageddon, which was the issue of deforestation,' Silva said. 'Radical deforestation.'

But first the new Environment Ministry had to commit to resolving it – and that meant a political commitment not everybody was prepared to make. 'There was a lot of resistance from people who thought it was fine to assume this commitment but it shouldn't be said publicly, because if it wasn't achieved an image of failure would be left,' Capobianco said. Silva was adamant. 'We can do everything that we think we have control over, and do nothing about deforestation, and nothing we did will be at all important,' she told one meeting. She won a commitment. The Action Plan for Deforestation Prevention and Control in the Legal Amazon (PPCDAm plan) put the environmentalists in charge, working with public servants and police, with political support from the highest power in the land – the president – and regular consultation with civil society. Environment agency Ibama was restaffed with new recruits, while existing employees were given more training.

Juliana Simões was already working at the environment ministry at that point. She recalled how many Ibama staff did not even have a high school education. A long-serving environment ministry official with a balanced and considered take on her subject, Simões left the ministry after Bolsonaro won election and it became clear that everything was going to change for the worse. Back in 2004, she told me on a video call, she felt optimistic as the government bought laptops, 4x4s and GPS systems for Ibama officials. 'The concern was to strengthen Ibama's technical team to work more intelligently on command and control,' she said.

Within months of the new government taking office, bad news about deforestation made headlines. Devastating figures were leaked in June 2003, showing a terrifying 40 per cent rise in deforestation. The numbers were for the period 2001–2002 – Amazon deforestation data runs from mid-year to mid-year – so before Lula's government assumed office, but they helped Silva's case. She told Lula they had to own the problem and he issued a decree setting up a working group. The following March, the plan was launched with the target of a 'strong reduction' in Amazon deforestation and fires, a 'substantial decrease' of land grabbing and a 'sharp reduction' in the sale of illegal wood. And crucially, the plan was centralised: it involved 17 ministries and secretariats but was coordinated by the powerful office of the chief of staff, José Dirceu. This put it at the centre of government.

Two months later, Brazil launched the DETER system, which provided deforestation alerts from imagery from NASA satellites. Previously, environment officials only had annual deforestation figures to go on, from a more detailed satellite system called PRODES which began operation in 1988, but whose results took a year to come out. By 2004, the results were coming out more quickly. Although less sensitive in capturing imagery, DETER began providing alerts of deforestation within two weeks and by 2011 those alerts were daily. Environment agency Ibama set up a special monitoring centre, enabling their teams working with the army and federal police to react rapidly. This was 'fundamental to reduce deforestation in the short term', said Juliana Simões.

In 2005, 480 police officers carried out 124 arrests and search and seizures in a showcase operation called Curupira, targeting a nationwide logging network that also involved middlemen and employees of both Ibama and state environment agencies. In 2019, 14 years later, the court case was still rumbling on, but that didn't matter, João Paulo Capobianco told me. Even without prison sentences, Operation Curupira still acted as a potent deterrent, he argued.

The plan's actions included monitoring, environmental control, and territorial management. Annual seminars discussed results and strategy. Dozens of new protected forest reserves, conservation units and areas where only sustainable practices like nut extraction and rubber tapping could be carried out were set up.

When deforestation began rising again in 2007 because, Silva said, offenders had worked out how to get around the rules, the government raised its game again. A new environment agency called the Chico Mendes Institute for Biodiversity Conservation, known as ICMBio, was created to protect conservation areas. A list of the municipalities with the highest deforestation rates was drawn up and publicised, with those on it suddenly finding their access to federal money and projects restricted. Credit for farmers was only approved if they could show they complied with environmental rules.

But by 2008, the political winds had changed direction. Lula was on his second mandate and his administration was being swayed by two influential lobbies – one pushing lucrative Amazon hydroelectric policies, another defending the interests of the powerful agribusiness sector. Matters came to a head during a brief meeting of Amazon state governors, held just before a ceremony to launch a 'Sustainable Amazon Plan' that Silva and her team had spent two years developing and which she believed she would helm.

But during the meeting, Amazon governors pressured Lula to revoke a crucial 2007 decree, numbered 6231, which had, among other things, created the 'dirty list' of highest-deforesting municipalities. According to Capobianco, Lula agreed to do so. And he announced that a Harvard-educated politician called Roberto Mangabeira Unger would head up the 'Sustainable Amazon Plan' – not Marina Silva. Lula told his environment minister she was 'from the Amazon and would not have the *samba* [a form of dance, perhaps used here to denote flexible movement] to get on with people', she recalled. 'It was in that context that I realised the commitment to do public policy the way it should be done was broken.' She resigned, believing that the repercussion this would cause would make it impossible for Lula to revoke the decree.

He did leave it standing, but a year later removed some of its clauses in a second measure.

Marina Silva was replaced by another environmentalist politician, Carlos Minc, who also kept deforestation down. In 2012, the Brazilian Congress approved a new Forest Code, which gave anyone who had deforested before 2008 an amnesty. This was regarded as disastrous by environmentalists and it is widely used as a defence by Amazon farmers, as I saw up close. As one environment official joked bitterly to me: 'Nobody ever cut down a tree after 2008.' Even so, the then president Dilma Rousseff sanctioned the code and, two years later, made Kátia Abreu, a senator and former head of the Agriculture and Cattle Ranching Confederation once dubbed 'the Chainsaw Queen', her agriculture minister.

Rousseff believes in using the state to develop Brazil's economy and, as a minister under Lula, had headed an aggressive programme of publicly funded works. And while Brazil signed up to the Paris Agreement on climate change in 2016, under her presidency, her environmental messages were confused. She aggressively pushed the mammoth Belo Monte hydroelectric dam project on the Amazon's Xingu River – savaged as an economic and environmental disaster. She battered Marina Silva into third place in the 2014 election with an aggressive smear campaign that still irritates many of Silva's supporters today, then narrowly beat conservative Aécio Neves to win a second term. The message that the environment would not hold up Brazil's development was loud and clear. Deforestation began to rise.

For the enforcers who fight environmental crime, the biggest problem is not simply the loggers, the land grabbers or the wildcat miners. Often, it's the people who might be expected to take a leading, responsible role but instead do the opposite: the local politicians, the environment minister, even the president. 'In Brazil, environmental crime is crime that pays,' said Antonio de Oliveira, a retired federal police officer in his mid-fifties, who regularly comes out of retirement for Amazon enforcement missions. We talked on the phone. He had just flown home from a mission with Ibama officers to blow up *garimpeiro* equipment on the western side of the Amazon, two months after being shot in the hand in an ambush by loggers while helping an Indigenous brigade called the Forest Guardians on its eastern reaches – which is where I had met him, when writing about them in 2015. He was a serious man with a military bearing, who carried a machine gun in the back of our 4x4 and simmered with indignation about the impunity criminals in the Amazon could expect.

'It is a political issue because those who finance a good part of the politicians, be they federal, state or municipal, connive with those who foment environmental crime,' Oliveira said. 'People who should be there to protect, to find resources and do good for the environment, are more and more doing the opposite.'

This frustration that environment enforcers feel about politicians is widespread, as I discovered in Brazil's brutal, modernist capital Brasília. Like many of the most striking buildings here, the Bank of Brazil's Cultural Centre was designed by the great modernist architect Oscar Niemeyer and its futuristic curves, neatly lawned gardens and terrace café make it one of the most agreeable places in the capital. It seemed like a good place to meet José Morelli, an Ibama official with a story to tell about how politics and environmental protection collide in Brazil.

Laying Down the Law: Political Leadership

An irreverent, laconic man in his late fifties, with a tangled and bitter sense of humour and a smoker's cackle, he joined the agency in 2002. Above the hubbub of the busy café, Morelli explained how he had become unwittingly famous in Brazil for inadvertently crossing someone who would become a very powerful politician.

Morelli was in charge of the Ibama office in Angra dos Reis, a busy tourist town on the Rio de Janeiro coast, during an operation in 2012, heading out in boats every day to patrol protected marine reserves. On 25 January, he and colleagues were on the water in the Tamoios Ecological Station, a federal conservation area where visits, diving, fishing and pretty much anything else is prohibited, when they came across a boat. There were three people in it, with fishing rods and buckets of fish, but when Morelli told them they were in a prohibited area and had to leave, one man very quickly got angry and insisted he had a letter from 'the minister' giving him permission to fish there.

It was then Morelli realised he was dealing with a notoriously combative, ultra conservative congressman called Jair Bolsonaro.

Undeterred, he told Bolsonaro that even a letter from the minister would not allow him to fish there. He would have to leave. 'No, I'm not going because I'm only fishing for small fish and I'm not fishing to make a living,' Bolsonaro told him. The congressman began to get angry and used his cellphone to call the minister of fishing, Luiz Sérgio de Oliveira, a former unionist from the Workers' Party, then in power. When that didn't work, Bolsonaro began to accuse Morelli of belonging to the same leftist Workers' Party. But faced with six armed Ibama agents and the implacable Morelli, Bolsonaro eventually sailed away, vowing to return the next day.

HOW TO SAVE THE AMAZON

On 6 March, Morelli fined Bolsonaro 10,000 Brazilian reais, then worth US$5,700, small change for a federal deputy. 'I didn't plan to, when I got there,' said Morelli. 'He takes everything dead seriously. That's his temperament, especially if he feels confronted.' The tortoise-like journey of this fine through Brazil's legal quagmire, and Bolsonaro's refusal to pay it, is emblematic of how powerful Brazilians ignore environmental law and get away with it.

Bolsonaro initially argued he had not been on the boat on the day the fine was issued, which was misleadingly true because the fine came after the transgression. But he could not escape. Morelli had taken a photo of him with a fishing rod clearly visible behind him. Ibama confirmed the fine. Bolsonaro then presented a completely different defence, complaining that he had been the only one fined and questioning why he and his two companions had not been arrested on the day. His defence also argued that amateur fishing or sport fishing was actually allowed. That appeal too was rejected and the fine confirmed in October 2013. That same year, Bolsonaro presented a bill to Congress to revoke the right of agents for Ibama and ICMBio to carry firearms, in spite of the danger they regularly face. Brazil's *Folha de S.Paulo* newspaper reported in 2018 that during one Congressional hearing, Bolsonaro told Ibama employees he would drop the bill if his fine could be resolved. In June 2015, he withdrew his bill. Two days later, the Attorney General's office argued his fine should go back to the very beginning of the legal process, arguing he had not been able to appeal – which he had. The fine remained in a state of suspended animation for the next three years.

Meanwhile, Brazil was in political turmoil. The ruling Workers' Party had been beset by Brazil's worst recession in decades and a sweeping investigation called Operation Car Wash had revealed widespread graft at state-run oil company Petrobras that saw executives, middlemen and politicians from the Workers' Party and its Congress allies imprisoned. Even though she was not accused of any crime over the scandal, Workers' Party president Dilma Rousseff became a target for anger. Millions of conservative Brazilians wearing the soccer team's national colours filled the streets of major cities to demand her ouster. In August 2016, she was impeached, ostensibly for breaking budget rules, following a torturous, eight-month process.

By 2018, Brazilian justice was as turbulent as its politics and the two were intrinsically linked. Sergio Moro, a hardline judge who had become a hero to conservative Brazilians for jailing many during Operation Car Wash, had imprisoned Lula over a seaside apartment that prosecutors claimed had been renovated for him by a construction company benefiting from government contracts, but which Lula argued he had never owned. The sentence was confirmed, removing Lula from an electoral race he was polling to win. By September 2018, with Lula out of the way, Bolsonaro was stabbed in the abdomen by a lone maniac during campaigning. From then on, he ignored debates and talked directly with supporters via his popular social media accounts. Morelli took a look at the status of Bolsonaro's fine and realised that it was close to the date it would expire. He alerted Ibama's president. 'I said, "President, that Bolsonaro fine is stuck, deliberately, and if it is not judged, it will expire."' Ibama in Brasília then told its Rio office to get the fine paid.

Bolsonaro won the presidential election comfortably in a run-off vote. He made Moro his justice minister, which for many confirmed a widely held belief that he had acted politically and for Bolsonaro's benefit in imprisoning Lula. Then, in June 2019, investigative site The Intercept Brasil published cellphone chats between Moro and Operation Car Wash prosecutors, obtained by a hacker, that clearly revealed the judge had colluded with the prosecutors. The credibility of both judge and operation was shattered. Lula was released and his sentence later quashed by the Supreme Court, making him free to run again. In 2020, Moro quit the Bolsonaro government, alleging that the president had interfered in the federal police. Things had gone full circle and the disgraced former judge was now accusing his former boss of interference, just as Moro himself was accused of having interfered in Lula's case.

With Operation Car Wash discredited, Brazil's legal system was left as dysfunctional as ever. And that, ultimately, benefited politicians like Bolsonaro, whose illegal fishing fine sat in a legal quagmire for another year, until October 2018, when Ibama's head office in the capital Brasília said it should be paid. But that same month, Bolsonaro won the election, and in January 2019, after his government took over, his environment minister Ricardo Salles began replacing regional Ibama bosses with military officers. In Rio, a reserve rear admiral took over the Ibama office, which decided that Bolsonaro's fine, finally, had expired. It was never paid.

On 25 January 2019, a tailings dam – which had held back a huge volume of toxic slurry and liquid waste – collapsed at a mine run by the minerals giant Vale near Brumadinho in Minas Gerais state. The torrent killed 272 people. Photographer Nicoló Lanfranchi and I were on the Amazon's northern border and we raced south to cover the horrific disaster. Every time politicians talk about building more mines in the Amazon, I recall the grief, the rescue workers buzzing helicopters and the stench at ground zero, where hundreds of mine workers' bodies, cattle and a pond full of fish were buried under a sea of sticky red mud.

Morelli was running Ibama's air operations unit that day, as the agency mobilised to deal with tens of millions of litres of mining waste that had flooded across rolling hills, burying a luxury guesthouse and its guests. This was the second disaster of this kind in recent years – another tailings dam, at Mariana in the same state, owned by a company called Samarco, had collapsed killing 19 in 2015 – and the river of waste caused an environmental disaster. Samarco was a joint venture between Vale and Anglo-Australian multinational BHP Billiton.

The Brumadinho disaster was a major environmental emergency, and yet three hours after the dam broke, Morelli was pulled out of an emergency meeting by an Ibama boss and told he was being demoted.

'Man, are you crazy? We are in the middle of a crisis,' Morelli told him. The agency backed off. But not for long. Two months later he was demoted – he couldn't be sacked because he had passed the civil service exam. Now he looks after a plane that monitors environmental accidents, a much less senior role. 'To be punished for doing what you are supposed to do is a disincentive, right? You say, "Wow, it's better not to do anything,"' he told me bitterly. 'It was demoralising for Ibama because the agency submitted to it and accepted it.'

After 20 years in Ibama, Morelli has very clear ideas about the deeper causes of Amazon deforestation and how it can be stopped. 'If you took the 50 biggest crop and livestock farming groups operating in the Amazon and did an environmental audit, especially in the areas they rent,' he said, 'you practically reduce deforestation in the Amazon by 90 per cent. Because these are the companies that finance deforestation in the Amazon.' As for illegal wildcat mining, destroying diggers and barges and closing illegal pits is futile. 'Who is the guy that financed that machine worth half a million real? That guy is the owner of all this. You never get to this guy,' he said. 'Why don't these owners of the machines go to Brasília, and present themselves?' Morelli asked. 'Because they are all crooks.'

How much of the Amazon hardwood sold around the world that claims to have been sustainably and legally produced really has? According to a study in the state of Pará in 2017–2018 by the Belém-based NGO Imazon, 24 per cent of permits for forest exploration showed serious 'inconsistencies' and other problems, while 70 per cent of the wood that was removed from state forests had no licence at all. The federal police officer behind Brazil's biggest ever seizure of Amazon wood thinks the situation is even worse than that: his off-the-cuff estimate was that just 1 per cent of Amazon hardwood was legally produced, a figure wood companies would undoubtedly contest.

I met Alexandre Saraiva in 2018, when Brazilian journalist Nayara Felizardo and I turned up at federal police headquarters in Manaus, and Felizardo talked our way into his office. The federal police is Brazil's equivalent of the FBI and it also controls the country's borders. Back in 2018, Saraiva was the federal police chief for the vast state of Amazonas. His was a dangerous and unforgiving role, as his colleagues battled rising environmental crime and vicious drug gangs in a region of few roads and endless rivers. Threats came with the job. He reluctantly received us from behind his big desk, leaned back in his sharp suit and Clark Kent glasses and agreed to answer our questions.

By 2021, he had become one of the most famous police officers in Brazil, appearing on a leading political interview show and being quizzed at Congress. After running the operation that led to Brazil's record wood haul, Saraiva had come under attack from then-environment minister Ricardo Salles, who had been nicknamed 'Minister for Deforestation' for his slavish obedience to Bolsonaro's destructive Amazon agenda. Salles flew to the Amazon to defend the logging companies. Saraiva denounced him to the Supreme Court – and lost his job. After ten years spent running federal police operations in three Amazon states, when we talked he was now running a federal police base in a small town in his home state of Rio de Janeiro. He did not appear to harbour any regrets, had grown a substantial beard and on his forearm, a tattoo no longer hidden under a suit read: 'BORN TO WAR'.

'I don't run from trouble,' Saraiva said. 'Never. My wife was saying this yesterday: "You don't give up, huh?" I said: "No, I'm going to carry on until . . . while I'm healthy and strong, I won't stop."' He said the involvement of powerful and important politicians in Brazil enabled environmental crime to rage as fiercely as it does. By 2022, he was planning to run for Congress for a centre-left party.

Saraiva had a lot to say about political interference in Amazon law enforcement. After one successful federal police operation, he recalled being called into a meeting with a powerful, nationally known politician who I won't name for legal reasons. The politician told Saraiva to abandon one investigation because the Amazon state in question needed the income produced by the illegal activity the operation had taken aim at. 'Think about a well-mannered guy, someone smooth,' he said, describing the man. 'He reminded me of the myth of Lucifer. The devil is a very agreeable person.' This politician was 'very smart. And as dangerous as he was well-mannered.'

Saraiva also mingled with Jair Bolsonaro, and the two of them got on fine. While still head of federal police in Amazonas, he even appeared on one of the president's weekly Facebook Lives. After Bolsonaro won the presidential elections in 2018, Saraiva spent two and a half hours at Bolsonaro's house in Rio, being quizzed as a possible environment minister. 'He's a nice, agreeable person on a personal level,' Saraiva said. But he added: 'We disagreed about almost everything.'

Saraiva argued that protecting the environment is beyond any one political creed. It doesn't belong to the left or the right. 'Everybody breathes the same air, everybody drinks the same water,' he said. 'It can't be ideological.' That sounded like a very coherent argument to me.

The record-breaking police and army operation Saraiva led happened in December 2020 and seized 227,000 cubic metres of Amazon hardwoods in logs. Operation Handroanthus – whose title came from the scientific name for the Ipê tree – was spurred after officers flew over a barge heading up the Amazon's River Mamuru containing nearly 3,000 cubic metres of wood and became suspicious. When the federal police turned up to check it out, nobody claimed 70 per cent of the wood, while the documentation presented for the remainder by its owners was full of 'irregularities', Brazilian news magazine *Veja* reported.

It is possible to legally farm wood in the Amazon, provided the farmer has a 'forest management plan' which has been approved by the relevant state environment agency. In theory, that is where the 'sustainable' Amazon hardwood sold in Europe and the USA comes from. But as Imazon's report showed, and Greenpeace has detailed in a series of studies dating back to 2014, these 'management plans' are widely problematic. 'Due to various forms of fraud that are common at the licensing, harvesting and commercialisation stages of timber production, it is almost impossible to distinguish between legally and illegally logged timber,' Greenpeace said in a 2018 report.

Saraiva agreed. 'I've looked at over a thousand administrative processes that authorise deforestation and I haven't found one that was solid,' he said. 'The fraud is grotesque.'

The Environment Ministry initially celebrated his operation. Then powerful politicians took the side of the loggers and everything turned upside down. In February 2021, the army suddenly announced it would no longer guard the seized wood. In late March, the then environment minister Ricardo Salles went into battle on behalf of the loggers and flew to Santarém in Pará twice, posing for photos in front of the piles of apprehended wood and inviting the media to witness the logging companies handing over documents to the federal police which, he said, proved the wood was legal.

Salles was already one of the most rabid ideologues in the Bolsonaro government, a lawyer from São Paulo who became environment minister without ever having even been to the

Amazon. He had consistently attacked and demoralised environment agencies like Ibama that he was in charge of. He became yet more notorious after suggesting during a shambolic ministerial meeting in April 2020 – a video of which was later released by a Supreme Court judge – that the government take advantage of media attention on the Covid-19 pandemic to send a 'cattle drive' through Brazil's environmental legislation, 'changing all the rules and simplifying norms'.

Salles attacked Saraiva and his operation in an interview – an unprecedented move by an environment minister. He said he had dug into the documentation behind two of the logs seized and found everything in order. Saraiva kicked back with an interview of his own, defending the investigation. He then formally accused Salles in a 'crime notice' to Brazil's Supreme Court – the only court that can judge ministers – arguing he had obstructed federal police work and sponsored the 'private ... and illegitimate interests' of loggers. He alleged that Salles was part of the same criminal organisation that his investigation had targeted.

Saraiva was demoted the next day. 'Everyone in the police knows it was retaliation,' he told me. Weeks later, he took part in a tumultuous session in Congress, making a detailed presentation about why and how he believed the loggers involved in his operation were guilty of land grabbing, forging documents and illegal logging. In May, the drama took another sudden turn, when federal police carried out a search and seizure investigation at addresses linked to Salles – his home, his law firm – and those of other ministry employees, including Eduardo Bim, the Ibama president Salles had appointed, who was suspended. Salles quit weeks later.

In ordering the investigation into the environment minister, a Supreme Court judge said police had uncovered a 'serious scheme to facilitate the smuggling of forest products'. Fourteen months earlier, on 25 February 2020, Ibama president Eduardo Bim had changed the rules to allow Amazon wood to be exported without Ibama authorisation, at the request of two logging organisations. Before that, on 4 February, the Ibama boss in the Amazon state of Pará, a retired police colonel from São Paulo with no environmental agency experience called Walter Mendes Magalhães Júnior, had retrospectively issued licences for five containers containing 110 cubic metres of Amazon hardwoods being held at the US port of Savannah in Georgia by the US Fish and Wildlife Service because they had arrived without those Ibama authorisations. These acts by Salles' subordinates were used as evidence against him. It was the first time an environment minister in Brazil was accused of environmental crimes.

According to Federal Police, the changes had been requested by two groups of wood companies – the Pará Association of Wood Exporting Industries, AIMEX, and the Pará Association of Forest Concession Companies, CONFLORESTA. In a joint statement, both organisations said they had 'always acted in the defense of the interests of their associates and the forestry sector in a firm but absolutely honest, legitimate and democratic manner' and that 'the products exported, and which are the target of the investigation cannot, under any circumstances, be labeled as illegal or contraband. The entire process of licensing the cargos was duly authorised by the competent environmental agencies.'

Five of the companies in AIMEX and one of its directors have between them been handed millions of reals in environmental fines by Ibama. One of the accused is a subsidiary of a British company called Tradelink with an office in the Amazon state of Pará, which has has been fined several times since 2010. In a statement to journalists at the time, Tradelink said all of its operations 'were legal and followed IBAMA's rules and the environmental agency's interpretation of the relevant legislation'. The company also said that the police's operation was just 'an investigation, and none of the allegations have been proved'.

Governments need to impose stricter traceability requirements on wood products, and companies should be more transparent about their supply chains. That would enable timber consumers around the world, including in the UK, to do their due diligence, make sure that what they are proposing to buy meets all of the current national standards, and is not connected to companies whose activities have either been the subject of fines for environmental infractions or attracted socioenvironmental lawsuits or conflicts.

We drove into Anapu, through an arch that read: 'Welcome, you are our happiness.' Underneath the word Anapu, the letters CV had been spray-painted for the Comando Vermelho, or Red Command, one of Brazil's most powerful and vicious drug gangs. It was just one sign of how violent this dusty roadside town on the Trans-Amazon highway in Pará state can be. Then, we crossed a wooden bridge with rattling planks, and found a shrine to Sister Dorothy Stang, an American nun and activist who fought to get land for poor Amazonian people to live on and protect, who was murdered in the forest an hour from here in 2005, in a killing that shone the world's spotlight on the Amazon's relentless brutality.

Her group of Catholic nuns still lives and works here. And on their patch of land there is a small forest reserve named after her, a rudimentary chapel and a nursery where they grow seedlings for the poor farmers she was helping when she lost her life. There was no security on their compound, just a wooden gate. A tall, red wooden cross plunged into the ground beside the shrine had 19 names written on it, of local people killed since 2015 in the region's ruthless scramble for land.

'It's not just her,' said Sister Jane Dwyer, Stang's octogenarian colleague who was still working here with the poor and landless. 'It's everyone in this municipality, starting with us. Life here is threatened.'

The two gunmen who killed Stang one morning in the forest with six shots, the middleman who arranged it, the two ranchers who paid 50,000 reais to have her killed have all served prison time, if sporadically and never for very long, considering the gravity of the crime. One of the landowners, Vitalmiro de Moura – a.k.a. Bida – has been arrested, tried and released four times. Some of the chaotic legal processes were captured in the documentary *They Killed Sister Dorothy*, voiced by Martin Sheen. But at least something was done: for the other 19 victims, nobody has been tried and, in many cases, locals feel police barely bothered to investigate.

Stories like this just keep coming, like shots from an automatic rifle. They are as repetitive as the plots of old Western movies. The murders of peasants and smallholders and Indigenous leaders working or protecting a piece of forest, cut down by gunmen paid by ranchers and farmers to do their dirty work, are crimes that fizzle out in the quagmire of the Brazilian justice system and the damp indifference of local cops.

I sat down with Antonia Silva Lima, known as 'Tunica', then 65, an expansive, warm-hearted woman with tangled grey hair, to try to understand a little about life in a region with so much violence and tension, a place people warn journalists to take care in when visiting. And to glimpse some of the impunity that holds the Amazon back.

Tunica and her husband José lived in the Esperança – or Hope – Sustainable Development Project in the forest where Sister Dorothy was killed, but we met at the nuns' compound. Her life had been threatened and it was not safe for her to be seen talking to strangers like me where she lived. A well-known character in the forest community who was close to Sister Dorothy, Tunica and her husband had remained true to the conditions of living in PDS Esperança, where small farmers were allowed to farm 20 hectares of their 100-hectare plot and leave a large communal reserve corresponding to 80 per cent of each person's land intact – as is the law in the Brazilian Amazon. But not everybody felt the same way.

'People don't really value this 80 per cent reserve. It's not important to them,' Tunica said. 'What's important is deforesting and growing *capim* (grass for cattle).'

The federal government paid more attention to the settlement after Stang was murdered and a fragile peace held until 2015. INCRA, the federal land agency that officially controlled the settlement, had set up two guard posts but when it stopped manning them with local people in 2019, things began to fall apart and there had been unsolved killings.

In recent years, some of the settlement's residents had sold up even though, officially, that wasn't possible. Some residents cleared their land and sold the wood. Farmers bought some of the lots and joined them together to form bigger farms. Trucks carried valuable wood out at night. Tunica was used to cooking for health

Laying Down the Law: Political Leadership

teams visiting her hamlet and in March 2021 agreed to cook for an INCRA official on what she thought was a visit. But he turned up with around 40 people, she said, including police and environment officials, for an operation. People were fined for breaking environmental rules and a rumour began circulating that Tunica had been responsible for denouncing them to the authorities which, she insisted vehemently, she had never done. 'I don't know why they blamed me,' she said.

Some people began looking at her differently after the operation. Two men on motorbikes circulated the sparely populated forest, asking where Tunica lived. A prosecutor who had long followed events here advised her to leave, and she had spent three months outside the region. Tunica wanted to go back to her simple, cosy wooden house, even though her husband no longer let her sit out front on warm evenings. She loved it there, the land was good, she said, real 'Mother Earth'. 'Everything you plant grows,' she said. 'My God. It's very rich. But sincerely I'm really unhappy about this persecution . . . you know when you are living somewhere that you don't feel you're safe . . . that at any moment someone could come and attack you.'

She told me how she and her husband José had fled crushing poverty in the northeastern state of Maranhão and bounced from Amazon town to Amazon town, working for others on the land, battling to pay rent, bringing up their six children. In Itaituba, a *garimpo* town, her husband worked on farms and in mining pits. 'But he didn't get anything except illness and malaria,' she said, describing this endless journey as a 'pilgrimage'.

They were living in Rondon do Pará, nearly 500km away, when José heard there was land and well-paid work to be had in

Anapu. When he got there, he found out there wasn't, but Tunica and the family were packed and ready to move so they came anyway – only to find out that life was, if anything, even tougher. 'I arrived here suffering. We didn't have money to pay rent, and we were here struggling, struggling, struggling, until one day, this story of land appeared,' she said. Tunica asked who was involved in distributing this land along with INCRA and the answer came back: 'Sister Dorothy'.

Dorothy Stang, a nun from Dayton, Ohio, had been living in Brazil since 1966. She was part of the Pastoral Land Commission (Comissão Pastoral da Terra, CPT), a Catholic organisation that helps rural poor and campaigns against violence. During Brazil's military dictatorship (1964–1985), government policy was to populate and develop the Amazon, which the generals saw as a way of protecting this isolated region and providing work for people suffering drought in the country's poorest, semi-arid northeast region. 'The solution to two problems: men without land in the northeast and land without men in the Amazon,' said General Emílio Garrastazu Médici, Brazil's dictator at the time, in a 1970 speech that became a slogan.

As the Brazilian government worked to populate the Amazon, in this part of Pará INCRA gave out the better land to migrants from the south of Brazil. They were usually of European descent, whiter and seen as being better at business, so more likely to make their farms a commercial success, as Noemia Miyasaka, a professor at the Federal University of Pará specialising in family farming, told me. 'People from the northeast and the north could live near bigger farms, and serve as a labour reserve. Because they just have little plots, little plots of manioc, small subsistence plots,' she said. People from the northeast were more likely to be Black or mixed race. Such institutional racism is common in Brazil, where 56 per cent of the population is Black or mixed race.

Agrarian reform is also a long-standing demand in Brazil, where 1 per cent of the population owns 47 per cent of all real estate. In law, farms that are not producing can be seized by the government and divided up for peasant farmers, and since 1987, thousands of such 'Settlement Projects' (Projetos de Assentamento, or PAs) have been set up in Brazil – hundreds in this state of Pará alone. But too often people were dumped on land that had already been heavily deforested or put into forest but given no help or advice on how to protect it. Instead, often they were encouraged by INCRA to clear the trees and make it productive in order to gain title. So, in 1999, the government created a new model of settlement project, the Sustainable Development Project, or PDS – and that's the model Sister Dorothy was working on with landless peasants in Anapu.

Tunica's husband José ended up joining up with other landless peasants and occupying the land that became PDS Esperança. The problem was that older land contracts, which in theory had been cancelled, still existed in the minds of powerful ranchers who claimed the same land was theirs. While Stang and the nuns worked with the settlers to make the project work, a horrific, perennial Amazon story of landowners versus peasants began to play out. For years before she was murdered, Sister Dorothy received threats. She even turned down police protection because she insisted the people in PDS Esperança get the same security detail.

'There were times when she said, "I'm going to give up,"' Tunica said. 'She was crying, she lowered her head because of the struggle and the persecution.' Tunica's eyes welled up with sorrow; 16 years later the memory still stung. She sniffed and cleared her throat. 'She lifted her head and said, "No, I won't give up. I will go until the end. I will fight to the end for the land for my people." And she really did fight until the end, until her life was gone.'

Anapu was an eerie, grubby town on this Monday night. Its streets were lined with gas stations, rough-hewn construction shops and a ritzy cowboy clobber boutique called Belarmino's Country. A handwritten note behind reception in our hotel offered land for sale and a WhatsApp number. I whacked a spider the size of my palm in the shower with my flip flop. We ate in a little restaurant, trying not to catch the suspicious eyes of other diners. The next morning, we drove out to PDS Esperança.

Gentle, cheerful and unassumingly charismatic, Jane Dwyer wore a T-shirt and a skirt and a smile with steel behind it. Born near Boston in the United States, she had studied sociology and in 1963, she was on the avenue in Washington DC when Martin Luther King gave his 'I have a dream' speech. Later, she joined the Catholic Church, working in Washington's poorer neighbourhoods. She came to Brazil in 1972 and to Anapu in 1999.

'I came to where I wanted to be, in the midst of the people. Where I am until today,' she said. 'We go from community to community, meet with people, see what people want and need and try to help and sort it out,' she said. 'Nobody will defend these people.' PDS Esperança, and another PDS called Virola-Jatobá, were set up in law after Sister Dorothy was murdered. At Virola-Jatobá, the settlers did not organise their farming; instead, they made a deal with a private logging company. But that came to an end in 2012 and five years later hundreds of armed squatters appeared one day to steal their land. It took almost a year for prosecutors and police to evict them. Ten days later, the invaders returned. Trucks rumbled out of the settlement carrying wood. It took another 18 months for police to act. It was still not enough: the logging continues. Sister Jane feels that if the government doesn't enter 'with claws' in both projects, the forest there is doomed.

The PDS 'was a project adapted for the Amazon in which the poor were on the richest land in the region, in the middle of the forest', she said. 'And the government supported it. Today, it's not like that.'

Government support fell off at the end of Dilma Rousseff's second government, when her former vice-president Michel Temer took over, she said. Politically, this was a point when the agribusiness lobby strengthened its hand. Nobody was ever going to make a fortune farming the small lots in Esperança, but that was never the point. 'You manage a decent, dignified, comfortable human condition, but you are not rich because it's only 20 hectares of land,' Sister Jane said.

Sister Jane is a Brazilian citizen. And after so many years here, she felt more comfortable talking in her heavily accented Portuguese than English. So that's what we did. She and fellow nun Katia lived in a simple wooden house behind a rickety picket fence, on a side street in Anapu. There were pictures of Sister Dorothy captioned 'mystic martyr, voice of the Amazon', Nelson Mandela and the pope on the walls. The two nuns sat in wicker armchairs and slept in hammocks in rooms out back. As traffic rumbled by, Sister Jane listed some of the settlements they worked with – and in all of them, the small farmers and their land were being besieged or faced legal battles or a community leader had been forced into hiding after threats to their life. 'They are being threatened and having their lots sold from under their feet,' she said.

Don't you ever feel like leaving? I asked her. 'But the people can't leave, so what right do I have to leave? I will leave when I die and can't help any more,' she said. 'Anyone who works like us is in the same situation. It's in every corner, it's not just us. Wherever the people are trying to hang on, this is the reality.'

Earlier that day, we had followed the 4x4 pickup truck with Sister Jane in it for an hour, over bumpy dirt roads into the Esperança land. We dropped Tunica off without getting out of the car or opening the window. At one point, another resident of the PDS, a farmer, agreed to talk to me if I withheld their name. Everybody in this rural area knows who the loggers are, who the powerful farmers are, who the hired gunmen are. Nobody expects justice to ever be done for the 19 people murdered from 2015 to 2019. 'Pará is considered a land without law,' the farmer told me once we had found a place by a stream where nobody could see or hear us. 'The law here is power and money, the law of the strongest.' This farmer had lost a family member, one of those killed.

'There are those who deforest because they need to, for a subsistence plot,' they said. Others 'go past the limits'. 'Many people come from the south, looking to take wood and land,' the farmer said. 'If you deforest, you never see the forest the same way again,' they added, listing all the uses of the different plants and woods – the superfood açaí berries, the oil from andiroba trees that works as a repellent. The farmer complained that those who kill rarely serve much time in prison, if at all.

Around us, the creek bubbled and wind rustled the trees. A relative appeared and proudly showed me a cellphone video of a spring, its water so clear that I could make out grains of sand on its bed. Such a simple, beautiful thing, threatened by guns, short-term greed, wood sold cheap and skinny cattle. By the same lawlessness that Sister Dorothy and many others have faced so bravely. As we walked back to the car, the farmer's words rang in my head: 'There are so many laws. If they worked, our Brazil would be an enchanted paradise.'

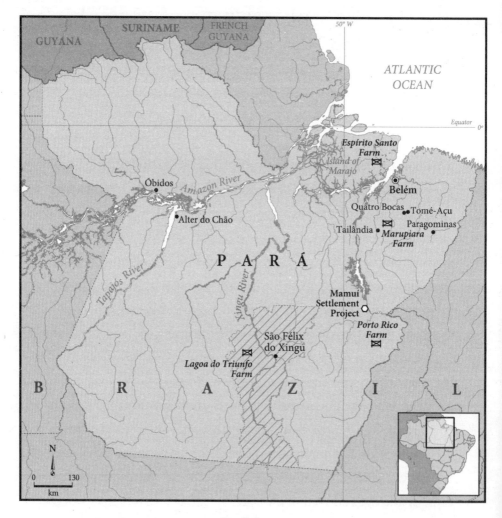

Pará State

CHAPTER TWO

Cattle Chaos: Corporate Accountability

Dom Phillips,
São Félix do Xingu, Pará

'The facts pointed out do not correspond to the standards and processes adopted by the company.'

We took the ferry across the Xingu River at dawn, drinking tiny plastic cups of coffee as its waters reflected the gold, lavender and turquoise lights of the sky. The curved sliver of a crescent moon glinted. Faces set in smiles in the early light, but Daniel Camargos, a veteran reporter from Brazilian investigative site Repórter Brasil, and photographer João Laet, who used to live in the region, and I all knew it was going to be a long, tense day. Investigating cattle farms in the Amazon is no easy feat, especially in São Félix do Xingu, a sprawling municipality of 84,000 square kilometres that is bigger than South Carolina and has a reputation for violence.

On the other side of the river, after bouncing down the ferry's rusty ramp in a rented 4x4 pickup truck and hitting a dirt road, we stopped for a rudimentary yet delicious breakfast at what was basically a café in a family house. We ate on the front porch – tapioca and eggs cooked in the open-air kitchen to one side, where pans bubbled on the traditional wood-fired stone oven. Then we drove off to find the cows, the skinny, white, rangy beasts that are at the heart of Amazon deforestation.

São Félix do Xingu is an unassuming, busy little riverside town at the end of a potholed highway. But on that trip in 2019, there were clear signs that there was money in this town. An upscale boutique called Villa Store had a handbag on sale for R$2,400 – US$609. Farmers' wives were among the store's patrons, sales assistant Kelli Moraes said, and the bag sold well. 'They are very fashion conscious. They love our brand, the more sophisticated clothes.'

It was a Saturday and we found ourselves at a two-day race meeting. Smart new pickups were parked up near marquees selling food and drink near a narrow race track. Women in tight jeans and boots and men in cowboy hats roared at the horses, slapping hands and waving wads of cash in jubilation when they won. Drinks flowed, music and horses' hooves pounded. The atmosphere was intense. One man with beer on his breath and a thick gold chain around his neck had bet over R$1,000 on one race, and won over R$1,300 on another, he shouted to me, over the noise of the sound truck.

Valdiron Bueno, the owner of two farming supply shops, was running the races. He told me there was over R$35,000 in prizes over the two days. He had moved to São Félix 20 years earlier. 'I came with just the clothes on my back and a bag like this,' he said, pointing to my small rucksack. We talked to two of the most important men in town: Arlindo Rosa, president of its rural producers' union – who arrived in 1993 – and his vice-president, Francisco Torres, who came in 1987. 'When I arrived here, there was practically none of this cattle ranching. This here was just forest, there was no road, there was nothing. There was no electricity, there was no communication,' Rosa said. Cattle ranching and hard work had built the town. 'People came from outside with the spirit to raise cattle.'

Now, there are hundreds of farms. This little town has the country's biggest cattle herd, with 2.4 million head of cattle for its 136,000 people, more than 17 per person.

Like many regions in this 'deforestation arc', the town also has a history of land grabbing and murder. One local man told us that when a local heavy came looking to buy up his land, he was told: 'Either you sell to me or your widow will.' According to the Pastoral Land Commission, in the last four decades, 62 workers and leaders have been killed over land disputes in São Félix do Xingu and nobody was ever found guilty. A family of three who lived in a remote waterside house and was known for releasing baby turtles into the river was butchered in January 2022.

It is cattle ranching that primarily drives deforestation across the Amazon. Researchers at MapBiomas, a non-profit group producing maps and data on the region, say that 88 per cent of Amazon areas deforested over the last three decades became cattle pasture. Sixteen per cent of the Brazilian Amazon is now dedicated to ranching and 42 per cent of Brazil's entire cattle herd grazes in Amazon states.

Multinational Brazilian meat companies like JBS, Marfrig and Minerva have been repeatedly shown to buy cattle from farms that have illegally deforested land. Despite long-standing commitments and millions of dollars invested in monitoring systems that are checked by independent auditors, these companies are unable to control their supply chains. For decades, farmers and cattle dealers have routinely and openly been moving cattle around farms in a series of movements which cannot easily be tracked.

How did this happen?

We drove for hours on a bumpy, red-dirt, public road across an enormous cattle ranch called Lagoa do Triunfo. It takes up 145,000 hectares, making it twice the size of New York City. The farm is owned by a company called Agropecuária Santa Barbara, or AgroSB, controlled by Daniel Dantas, a controversial billionaire once described by the business site Bloomberg as the 'bad boy' of Brazilian finance, who bought up half a million hectares of the state of Pará for cattle ranching after becoming embroiled in a corruption scandal in 2008. He was arrested, briefly imprisoned, and his fund forced to give up some of his assets. But here in the south of Pará, his cattle business was booming, despite the fact that between 2010 and 2013, environment agency Ibama had fined AgroSB more than R$18 million for deforestation on the farm. At least 12 areas of land on the Lagoa do Triunfo farm had been embargoed by Brazil's environment agency Ibama – meaning they could not be used.

That meant that a responsible meat company like JBS – the world's biggest meat packer – should not have bought any cattle from this farm. And this is where it gets complicated. Because it is commonplace for cattle to be moved from farm to farm. In Brazil, some farms breed cattle, others then buy those cattle to fatten and different farms then buy those cattle to 'finish' and sell on to slaughterhouses. The monitoring systems used by meat companies only trace the 'direct' suppliers – the last farm in the chain which sells to the slaughterhouse. JBS, Marfrig and Minerva originally promised to establish control over the entire chain by 2011, following deals with federal prosecutors. But all three companies were unable to meet all of their publicly stated goals.

Cattle Chaos: Corporate Accountability

We already knew from sanitary movement documents called GTA (Guia de Trânsito Animal – Animal Transit Guides) used to ensure vaccinations and control diseases like BSE, 'mad cow', that JBS had bought thousands of head of cattle from another farm called Espírito Santo, which is also owned by AgroSB in this same state of Pará. And we knew that AgroSB had also moved nearly 300 cattle from Lagoa do Triunfo to Espírito Santo. We also knew that in 2018, Santa Bárbara had sent over 700 head of cattle from the Lagoa do Triunfo farm to its Porto Rico farm, which in turn had supplied dozens of cattle to another JBS slaughterhouse. What we had come to see were the cows on the Lagoa do Triunfo's embargoed land.

There were scattered patches of trees on distant hills, but little sign of forest, even though this entire region was – on paper at least – a protected reserve. The car lurched and bounced on the dusty road. We left the farm's limits, drove to a store, ate snacks, got caught up in a big herd of cows being moved along the dirt road, then headed back over a wooden bridge to recross Lagoa do Triunfo again. Then we saw a small group of cows on land where no farming was supposed to take place.

Hours of bumpy roads later, after dark fell, we talked to two young men who had rolled up to a small bar and shop in a hamlet just outside the ranch's borders. Both worked for AgroSB and lived on the ranch. I am withholding their names to avoid any repercussions for them. The work was good, well paid, they were treated properly, they said. One man said cattle were allowed to roam in areas employees knew were embargoed. 'You can't cut down the vegetation,' he said. 'The vegetation grows and we work the cattle inside.'

Agropecuária Santa Barbara said that any deforestation had happened before 2008, when it bought the farm. Coincidentally, this date is also the cut-off date for deforestation amnesties under Brazil's Forest Code.

'AgroSB does not carry out deforestation in order to increase its area, but rather it recovers degraded areas,' the company said in an email. 'AgroSB's business model is anchored in the acquisition of degraded open and pastured areas, which are fertilised, reclaimed and transformed into high-intensity pastures or grain plantations.'

JBS also denied it had broken its own commitments. 'The facts pointed out do not correspond to the standards and processes adopted by the company,' it said at the time, pointing to its extensive monitoring system and an independent audit from 2018 that found 'more than 99.9 per cent of JBS's cattle purchases meet the company's socio-environmental criteria'. However, the independent audit, from Norwegian company DNV GL, also said that: 'Indirect suppliers of cattle to JBS are not yet checked systematically, since JBS has not yet managed to adopt auditable procedures for its indirect suppliers.'

More recently, this vast global company has pledged to put in place a tracing system for both direct and indirect suppliers by 2025. If the technical and financial obstacles are overcome, that might close the loopholes that have long enabled the company to buy cattle that have at some stage been raised by farmers involved in deforestation, land grabbing and other environmental crimes. But the bigger challenge is corporate culture. JBS has profited from Amazon destruction for so long that it has much to prove if it wants to demonstrate that it really can be part of a solution. Past

Cattle Chaos: Corporate Accountability

international commitments to change tack have often not been born out by actions on the ground.

Arlindo Rosa and Francisco Torres, the president and vice-president of the rural producers' union who I had met at the racetrack, had been fiercely critical of environmental fines and protection measures, which they argued held up the region's development. 'There are too many reserves here,' said Rosa. 'More than 90 per cent of properties have embargoes or PRODES (a satellite alert of deforestation or fire).' That meant ranchers were unable to sell their cattle legitimately or apply for credit that might allow them to improve their productivity, allowing them to farm more on the pasture they have rather than deforesting additional land, Torres argued. 'I believe that the rural farmer, in the municipality of São Félix, already has roots here and needs land regulation, so he can get his land title, juridical security and bank credit to be able to invest in technology.'

This is a contentious argument that is widely used by Amazon ranchers, and I will examine it later in this chapter. But its persuasive power is lost when the ranchers using the argument rail against law enforcement, as both men continued to do. 'There are fines here that you can never pay,' Rosa said. 'This business of fining four million, five million (reais).'

None of this seemed to have impeded the São Félix cattle industry. Nor did Rosa mention that he has three embargoes of his own on the Ibama website and nearly five million reais in unpaid fines (around US$1.9 million at the time).

Two and a half years later, nothing had changed. A piece for Bloomberg exposed how cattle were still being sourced from farmers with fines and embargoes for deforestation. 'A ten day trip into the heart of Brazil's cattle country put on full display how easily and clearly cows from illegally cleared land flood supply chains,' reporter Jessica Brice wrote. JBS was still three years off its new deadline for establishing control of its cattle chain. The company declined to allow anybody to be interviewed for this book.

Just as in the United States, cowboy culture is a powerful force in Brazil. In the Amazon, it is more than just a way of life. It is a way of being. Men like Arlindo Rosa and Francisco Torres – most ranchers are male – see themselves as rugged individualists, frontiersmen, pioneers who achieved success through hard toil and bloody-minded determination. In recent years, Brazilian country music, called *sertanejo* and closely connected to cowboy culture, has overtaken more traditional forms like samba to become the country's most popular sound.

A 2017 study by nine international scientists on 600 households in this Amazon state of Pará found that even though cattle farming was not as lucrative, per hectare, as fruit farming, it was seen by many ranchers as more secure, both financially and socially. 'Cattle farming offers numerous perceived social advantages, including a quiet lifestyle, safety and social status,' two of the researchers wrote in an article about the study. As farmers often say, at times of need, they can always just raise cash quickly by selling a cow – not an option for those growing fruit or crops. Cattle ranching is seen as a respectable, desirable occupation. Within its close-knit social universe, Bolsonarismo is popular, and conspiracy theories about global warming – for instance, that it is a ploy by European and American farmers to take Amazon competition out of business – circulate.

In his book *Rainforest Cowboys*, anthropologist Jeffrey Hoelle describes the financial and cultural role that cattle ranching plays in the west Amazon state of Acre, where rubber tappers who also raise cattle and smallholders call cows 'living money' and 'savings'. Hoelle lived in Acre while researching his book and vividly explains how *cauboi* ('cowboy', given a Portuguese pronunciation) and *contri* (country music and style) became such dominant cultural and economic forces in the state. 'Although it never corresponds exactly with experience, cauboi culture is the only form of rural identity that is positively valued and thoroughly institutionalised throughout Brazil,' Hoelle wrote. 'Cauboi culture provides an oppositional voice to environmental preservation, regarding such conservation efforts as an affront to the self-reliance that is central to the identity of many rural producers, especially the migrants who came to Amazon to build their future.'

Ranching has also been heavily encouraged and financed by government. In 2007, the Lula government began buying heavily into meat and milk companies under a policy that set out to encourage the Brazilian economy through investment in 'national champions' – flagship capitalist firms which are selected to power industrial development. The government development bank BNDES injected billions into meat companies. In 2009, it put R$2.8 billion into JBS – a family butchers founded in 1953 – and invested so heavily in Marfrig that by 2016, the government owned a third of the company. In 2014, the government of Lula's successor, Dilma Rousseff, launched a plan called 'More Ranching' through the agriculture ministry, aimed at widening markets, improving genetics and bringing in more technology. 'Cattle ranching has a fundamental role in the development of the country,' an introduction to the plan read.

I had been travelling around the east Amazon state of Pará for two weeks when very early one morning I took a bus south from its capital, Belém. Many hours later, when I finally got to see a serious chunk of rainforest for the first time in this heavily developed region, it was inside a cattle ranch. Forest covers 80 per cent of Marupiara Farm, situated between the towns of Paragominas and Tailândia, owned by Mauro Lúcio Costa, a rancher with a very different approach to raising cattle profitably. Leaving 80 per cent of the land as forest is a legal requirement for most farms in the Amazon, but it is a requirement that less than a quarter of them meet, according to research commissioned by *piauí* magazine.

The 56-year-old Costa looked like the quintessential cattle rancher. He wore a cowboy hat and cowboy boots and his neat blue shirt with the Marupiara Farm logo on the pocket was tucked into cream jeans fastened with an enormous, gleaming belt buckle.

He rose at 4.30am every day to pray and walk for an hour before breakfast, he told me. He is a hard-headed businessman who crunches numbers, talks kilos of meat per hectare and profits, seeking to constantly improve his production. He is friends with many other ranchers and even voted for Bolsonaro in 2018. But Costa believes in protecting, not developing his land. He has reforested some of his land with precious hardwoods and has a more intensive and productive approach to farming cattle that some environmentalists believe could provide at least a temporary solution to the conundrum of how to protect the Amazon while allowing its hundreds of thousands of cattle farmers to continue producing meat. Essentially, he produces more meat on less land, unlike some ranchers who simply expand into nearby forest when their pasture is exhausted.

Costa is a tall, rangy man with a deadpan, mischievous sense of humour and he drove his white 4x4 pickup fast and efficiently. It is up to the government to monitor deforestation, Costa told me on the long drive out to his land. But it's also up to ranchers to produce more efficiently and sustainably – and stop blaming the government. 'In recent times, we've failed in our responsibilities,' he said. 'I am against deforestation and I do things against deforestation.' It's also up to ranchers who do produce sustainably to educate others. 'It's not just that I won't deforest, but that I will make people around me aware, I will work with these people so they don't either,' he said. 'We have to play a leading role.'

On backlands dirt roads, we passed a truck laden with logs – Paragominas has lost 45 per cent of its forest and there is not much left in this part of the Amazon. Costa talked the whole way, including about how he believes that biodiversity should be made a tangible asset that could help him sell beef. 'I look to take care of my forests. I enrich my forests and I plant trees in my forests. Because I believe in biodiversity,' he said. But he gets no sales benefits from this. He doesn't even know where his meat is sold – and whoever is buying it, whether in Brazil or the wider world, has no idea where it has come from.

Costa has a system of chips and tags that allows him to track where the cattle he buys have come from – and to calculate which of his suppliers provide the most profitable cattle, those that fatten up most. And he now has a cellphone app, he said, that lets him know if any supplier has environmental problems or has illegally deforested land. Much of the cattle he buys, he said, comes from a farm that births and raises cattle – but not all of it. And he can't tell if his suppliers have bought from a farm that has fines or embargoes.

That should be possible. 'Today, with the level of technology that we have, the resources and the rules we have, I think it is possible,' he said. But it's not helped by a 'tumultuous' relationship between ranchers and meat packers and what he calls a 'lack of self-responsibility'. The meat industry should find a way to resolve the problem. 'If this tracing process is in the rancher's hand, the rancher needs to do it,' he said. 'To say that this costs and the cost is unviable is a lie. Of course there is a cost, but this cost is very diluted and it really improves the management of your business. So this cost for the management of your business is negligible.'

Costa believes it should be possible to track the meat you buy anywhere in the world, by using a barcode, seeing where and how the cow in question was raised, seeing video of the farm – even having the opportunity to visit it. 'What I need to do is sell this product,' he said. 'I need to know what the market wants. So if I have a product that I produce with high biodiversity, with all the care, with animal wellbeing, all very pretty, why can't I go to the customer and say, "Look, this is what I've got, this is what I'm offering you."' He drew a comparison with ABS brakes, which manufacturers first included on more expensive cars until they became seen as standard. People are unwilling to leave their comfort zones, Costa said, and that includes ranchers. 'Normally, human beings are resistant to change.'

He grew up in the town of Governador Valadares, in the landlocked southeastern state of Minas Gerais, in a family of cattle ranchers. His father, who always owned farms, bought a 4,356-hectare farm with just 500 hectares of pasture in Pará in 1975. By 1978, almost half of it had been cleared. Mauro Lúcio moved here in 1982, to a 'very inhospitable place, with very little structure', he said. 'There was no electricity. There was very little. Communication was terrible.' And it was all forest.

Paragominas officials received regular deforestation reports from Brazilian NGO Imazon, which analyses satellite monitoring data. Mayor Adnan Demachki and his environment secretary Felipe Zagallo often drove out to farms where deforestation alerts had been spotted but often found it impossible to locate the culprit.

On 23 November 2008, at the request of Mayor Demachki, federal police confiscated 14 trucks piled high with illegal wood that loggers had taken from an Indigenous reserve. In retaliation, the offices of environment agency Ibama were set on fire – and Paragominas became national news. Demachki threatened to resign if the town didn't collectively sign up to a zero deforestation pact and a plan that became known as 'Green Municipality Paragominas'. This included ensuring everybody completed the rural register of properties, or CAR, which allowed local associations to know who was deforesting. The measures imposed a sense of collective responsibility that enabled Paragominas to meet the Environment Ministry's criteria and be removed from the Worst Municipalities list two years later.

Costa, then president of the rural producers' association, had supported the mayor's proposal. For him, the fact that farmers who legalised their land – by working with the authorities to satisfy regulatory requirements – were not immediately punished, that the town realised that deforestation in rural areas impacted everybody, and Mayor Demachki's ability to make this a collective problem were all important factors. 'He gave very, very good direction and one thing the mayor really achieved was the participation of society. He made society as a whole take part in the process,' Costa said.

Amazon cattle ranching is usually highly unproductive with farmers raising an average of 0.8 animals per hectare. Costa, by contrast, has an average of five animals per hectare on his farm and ten in some areas. His employees measure soil fertility each year and make adjustments as necessary by adding lime, phosphorous, potassium and nitrogen. These inputs help to improve the quality of the grasses. Associates say Costa has been fertilising the same area for 10 years and intends to continue this process for decades to come.

We turned off onto a track and the light went. No more occasional street lamps or lights in the odd wooden house; we were inside a dark forest. We drove through the forest for long while before we reached a spacious, cream-coloured, wooden farmhouse with a tiled roof and a long wooden chair on a terrace – a comfortable, spacious home. Costa disappeared to bed. Some of his cowboy hats hung on a stand in the living room, others on a screen on the terrace.

Early next morning, Costa took me to see his cows and his forests. He showed me the ear tags and chips that are used to count and trace cattle. One area of his ranch called Acácia had over ten head of cattle per hectare. He led me to one area of his farm where he said he is raising 8.19 head of cattle per hectare, about ten times the Amazon average. He also said he sells 80 per cent of his herd each year, compared to the average of 38 per cent.

Cattle Chaos: Corporate Accountability

A meat packing plant slaughtering 12,000 head of cattle a month would, he calculated, need 480,000 hectares of pasture to keep it supplied. At the quicker and more productive rate his farm works, the same plant would need just 45,000 hectares of pasture to keep it supplied. He waved at a plantation of African mahogany and led me into another area, where a mix of tree species were being grown. Later, he hopes to sell this wood. We passed a spider's web, a metre across, spread between the trees. 'This is biodiversity,' he said.

Later, a farm hand called José Silva, known as Santo, showed me an area of degraded forest where native hardwoods had been planted within the trees and bushes in straight lines. 'Look at the beautiful nature,' he said, pointing to a tortoise hidden inside its shell on the forest floor. 'There are jaguars but they don't mess with us.' Santo stopped by a skinny sapling. 'This here is an ipê,' he said, holding a narrow stalk that reached to his shoulder. 'In this area we have ipê, we have acapu, there behind you we have massaranduba. Further on we have jatobá. We have andiroba, we have parica.' Some of the trees were eight metres high. Santo pointed out where a band of peccaries had passed. 'They walk in a group – 20, 50 of them. You can see the ground is clearer.' In three to four years, the massaranduba had already reached five metres high.

We got back in the pickup and drove on, stopping by a wall of native forest. Santo found a way in and, as I tied my lace before venturing into the thick, tangled mix of plants, shrubs, trees and creepers, I heard the sound of something move suddenly nearby. Here the temperature dropped noticeably, despite the hot lunchtime sun. It was like walking off a scorched street into an air-conditioned building. As we pushed our way through the trees, Santo stopped when a tiny, green tiriba bird twittered; he gestured at the bigger, taller, more formidable trees around us. 'A logger would take all of these,' he said. We paused beside an estopeiro, stretching high above us. 'You are a little child next to her,' Santo said.

We had long since lost sight of the dirt road and pastures. It was remarkable how quickly we had become surrounded by thick forest as far as we could see. Santo pointed to a tree behind, twice as thick as him. 'This is a garana, that one next is a louro, louro vermelho, louro amarelo, the next one is called piquiá, behind us here,' he said turning around, 'sucupira, the other one there, acapu, and it goes on, there are more we can see.'

There was a chart in one of Costa's farm offices with the ranch's proportions clearly stated on it – 77.33 per cent of forest, 19.86 per cent of pasture, 2.03 per cent of riverbanks, 0.18 per cent being reforested, 0.6 per cent others, including a small area of soya. There were none of the signs of fire I had seen on other Amazon ranches – the burnt, dead trunks left standing or lying in fields of pasture.

Over lunch, Costa held forth with his wife Heloisa and two guests – one of whom was from one of Brazil's biggest banks. One of Costa's many plans includes franchising his ranching model to help other ranchers produce the same way. 'I hate fire. I only like it on a barbecue, to roast my meat,' he told the table.

Could it serve as a model for sustainable ranching? Yes and no. Fertilisers are not good for the environment. They cause more destruction when used carelessly. The run-off can cause eutrophication in water – a build-up of nutrients that leads to the rapid growth of microorganisms that use up oxygen, choking rivers and lakes. Studies have shown that the more cattle being raised on land, the higher the greenhouse gas emissions from their belches and farts. Nitrous oxide emissions are caused by use of nitrogen as a fertiliser. For many environmentalists, the idea that any cattle farming should be encouraged is heresy.

Added to this, Costa's message is capitalist, individualist, self-responsible. But while environmentalists might not agree with it any more than they like the fertilisers on Costa's farm, few would argue with the amount of thick, tangled forest he preserves. Given the desperate situation the Amazon finds itself in, in large part because of cattle ranching, the work of farmers like Mauro Lúcio Costa in trying to be more sustainable is welcomed by some Amazon environmentalists. They see his model as one of the ways to help save the Amazon, where cattle ranching employs 800,000 people. But others disagree, arguing that farmers like Costa are merely obeying existing laws and that the best conservation results would come from more protection, command and control.

To find out more about this side of the debate, I arranged to see one of the sharpest critics of ranching in the Amazon. Paulo Barreto suggested meeting at 6.15am at the gate of the Utinga State Park in Belém – the teeming, sweltering state capital of Pará. He was used to zipping around town on his bicycle and suggested a quiet, outdoor café overlooking a reservoir in this beautiful tropical park which Imazon, the NGO he works for, helped create. A tropical oasis flanked by favelas and busy highways, even at that time in the morning it was quickly filling up with runners and cyclists. I hired a bicycle and cycled alongside this balding, bearded and intense man, who talked as fast as he cycled. As we passed a freshwater lake, he pointed out fresh capybara shit on the asphalt as we went. At one point he gestured up to a noise in the trees. 'Monkeys,' he said.

Born in Bahia and raised in Maranhão and Castanhal, Barreto studied as a forest engineer and did a master's in forest politics and economy at Yale University. In 1990, he and four others founded Imazon – the Amazon Institute of Man and the Environment (Instituto do Homem e Meio Ambiente da Amazônia), which evaluates satellite data on deforestation produced by several space agencies, including NASA in the United States and INPE (the Brazilian National Institute for Space Research). The organisation works with communities across the Amazon on sustainable development and produces an important range of studies on issues such as land grabbing. Imazon is an important voice in the increasingly noisy conversation in Brazilian society about how to preserve the Amazon while allowing its population to improve its living conditions, not least because, unlike many other NGOs, it is actually headquartered in the region, here in Belém.

Barreto has written or worked on hundreds of studies. His argument – delivered in quick-fire, staccato bursts – is that it is impossible for Brazil to stop deforesting the Amazon while it encourages farmers legally and economically to keep doing so. He laid out examples of this for me, some of which are dealt with in his latest study, published in 2021: 'Policies to Develop Cattle Ranching in the Amazon Without Deforestation'. It was produced for a research initiative called Amazon 2030, studying the Amazon's economy and people, and suggesting action plans and solutions. Ranching is notoriously one of the Amazon's biggest problems but Barreto also believes it could offer some solutions.

Cattle Chaos: Corporate Accountability

'It is the main cause of deforestation but it is very inefficient,' Barreto said. He thinks that Mauro Lúcio Costa's farm – which he has also studied – offers some solutions.

Cattle ranching's takeover of large parts of the Amazon is a recent phenomenon. Iberian cattle first arrived in the Americas in 1494 and spread over North and South America during the next century. But it was only in the 1970s and 1980s that cattle began seriously impacting the Amazon. The move to this style of farming was led by large-scale ranchers and driven by land speculation, government subsidies and loans, as Brazil's military dictatorship sought to populate the region. 'If you look at the history of Brazil, ranching always had the role of opening frontiers,' Barreto said. This 'extractive ranching', as it is called, has proved the most efficient and destructive means of populating the Amazon and turning its lush forest into dry cattle pastures.

In 1975, less than 1 per cent of the Amazon had been deforested. In the 1980s, as hyperinflation roiled Brazil, cattle were seen as a cheap, safe investment. And putting cattle onto land has always been a good way to consolidate plots that had been squatted or simply stolen, as we had seen glaringly illustrated in the forests around Novo Progresso. Today, a fifth of the rainforest has gone, mostly turned into cattle pasture.

'The ensemble of benefits directly tied to the clearing of forest ostensibly for cattle ranching made this activity enormously attractive, and indeed the benefits were designed to lure investors and capital into the region,' American professor and Amazon expert Susanna B. Hecht wrote in 1993. The Superintendency for Amazon Development (Superintendência do Desenvolvimento da Amazônia, or SUDAM) gave out grants of up to 75 per cent of development costs. This especially benefited bigger ranches, responsible for 30 per cent of land clearances in areas where they dominated, such as northern Mato Grosso state and southern Pará state. Many of the farmers came from southern Brazil.

'The dynamic entrepreneurs from southern Brazil were given extraordinary favours, in part because they helped craft the terms of the incentives and because they were to take on the mission civilizatrice of taming the Amazon,' Hecht wrote. She described over 50,000 livestock operations. Thirty years later, there are 500,000.

Many of the most powerful ranchers in the Amazon are white, blue-eyed, and originally from southern or central Brazil. Their economic and political power derives from the racial colonisation of the Amazon. It began with attacks on, then genocide and enslavement of Indigenous people, and was followed by the forced immigration of enslaved Africans, who often survived better in the region than European colonials because they were more resistant to diseases like malaria – but often escaped to found the quilombola rural communities. Finally, as we have already seen in the case of Tunica and Marina Silva, migrants from the poorer northeastern states, themselves more likely to be Black or mixed race, had gone to the Amazon in the often futile hope of fleeing poverty.

White immigrants were given more chances. In the south of Pará, migrants from Rio Grande do Sul state were given better land – which many still own today. White Brazilians, who constitute 46 per cent of the population, are richer, healthier and have better access to education and services than the Black and mixed-race majority. This is markedly true in the Amazon, where the poorest people living in riverine communities or small plots of land on settlements are invariably Black or mixed race and the farmers with the biggest land very often are white.

According to Barreto's ranching study, Brazil's Agriculture Ministry has two projections for the growth of meat production 2020–2030 – of a conservative 1.4 per cent and more ambitious 2.4 per cent. Improving productivity to meet these targets demands a boost in productivity, through fertiliser inputs, land remediation, advanced breeding and other technologies. Without this, Brazil will need to deforest 634,000 to a million hectares a year by 2030 to meet demand. Baretto suggests helping ranchers raise cattle more productively on the land they already have, instead of clearing more. But the system needs to change to make this happen.

'The government says, "Look, you don't need to deforest," but they don't change the incentives,' Barreto said. His report argues that the government should prioritise rural credit for farms and municipalities that invest in improving productivity with public–private partnerships, anti-deforestation pacts, and land regularisation.

But he had a proviso. 'Cattle in terms of climate change is horrible. The methane emission produced for a unit of calorie and protein is eight times that of chicken,' he said. (According to the United Nations Food and Agriculture Organisation, beef emissions per kilo of protein are five times that of pork and eight times that of chicken.) 'So from the point of view of reducing climate change emissions, which cause climate change, we have to accelerate a transition to reduce the amount of beef.'

There is ongoing research into feed additives that would reduce cattle's enteric methane emissions – JBS is trialling one with 30,000 cattle, and some researchers believe a 30 per cent reduction in cattle emissions is certainly feasible. Instead, however, Barreto pointed to another potential solution: the rise in plant-based 'meat', including a slew of acquisitions of vegan food companies by meat giants like JBS and Marfrig. A 2021 study by the Boston

Consulting Group – a corporate consulting company – said by 2025 '11 per cent of all the meat, seafood, eggs, and dairy eaten around the globe is very likely to be alternative. With a push from regulators and step changes in technology, that number could reach 22% in 2035. By then, Europe and North America will have reached the point of "peak meat", and consumption of animal proteins will begin to decline.'

Vegetarianism has risen in Brazil – one study said 14 per cent of the population is vegetarian. But meat is deeply embedded in Brazilian culture and is also eaten widely in the Amazon – not just by cattle ranchers, but by riverine communities, forest peoples and Indigenous communities. Some hunt their own game meat, but not all of them. For example, the Raposa Serra do Sol Indigenous Territory on savannah land in Roraima state in the northern Amazon is home to 25,000 Indigenous residents from five peoples, many of whom raise cattle in small family or village groups. On a 2019 visit there for a meeting of leaders from across the reserve organised by the Indigenous association the Roraima Indigenous Council (Conselho Indigena de Roraima, CIR), we were fed beef for breakfast, lunch and dinner. Across Brazil, barbecues are part of family and social life. When a truck carrying carcasses of beef flipped on a highway near São Paulo in 2020, it was looted by crowds who carried away chunks of meat.

While they are paying very close attention to the rapid rise in sales of plant-based products and buying up plant food companies, meat giants like Marfrig don't see the beef market disappearing: 'A radical change in consumption habits is happening. Our children or young people will probably have consumption habits that are very different to what we have,' Paulo Pianez, Marfrig's director of sustainability and corporate communication told me. 'So our position today at Marfrig is that we are a protein company, which could be animal protein because there will be space for this type of food for many, many, many, many years,' he said.

Other Amazon environmentalists also advocate reforming ranching, rather than trying to work to eliminate or severely curtail it. Caetano Scannavino had for three decades been the coordinator of an Amazon NGO called Saúde e Alegria (Health and Happiness), based in Alter do Chão on River Tapajós in Pará. Its work includes health education, a hospital boat visiting remote riverine and Indigenous communities, an experimental agroforestry centre and even communication and tourism training for young Amazon residents. Caetano is an important voice in the Amazon conservation debate, and the NGO he runs with his brother Eugênio Scannavino, a doctor and infectious diseases specialist, has won several prestigious awards.

'I am not against agribusiness, I am against "ogre-business" – illegal deforestation, land grabbing, slave work and illegality. We have to reward those who are doing the right thing and punish those who are doing everything wrong, and not the reverse,' Caetano told me. 'The big villain in the Amazon is ranching, not soya, and if there are very unproductive pastures . . . why isn't there a policy of encouraging agricultural efficiency and increasing production without deforestation?'

A study by the non-profit group Imaflora examined the greenhouse gas emissions of five farms in Mato Grosso state who progressively adopted so-called 'sustainable ranching' practices as part of a project called the New Countryside Programme. The potential reduction in emissions of greenhouse gases reached

50 per cent per hectare and 90 per cent per kilogram of meat produced, it found.

But there is fierce debate around the issue, not least because of the wider ramifications. For example, others argue that encouraging more intensive cattle ranching could increase land values – prompting more land grabbing – and has little proven effect. A 2017 study by American natural resource economist Frank Merry and Brazilian professor of environmental modelling at the Federal University of Minas Gerais Britaldo Soares-Filho rejected the idea that helping cattle ranching intensify benefits the environment. They argued it was better to leave the cattle industry to do its own intensifying, which it will inevitably do, as that is where profit lies, and introduce stricter controls.

'The inescapable truth of beef production is that it is one of the least efficient transformations of energy into consumable calories, and whether driven by government policy, markets or custom, the increasing consumption of beef worldwide will undoubtedly have significant environmental impact,' Merry and Soares-Filho wrote. 'One should focus on causes, not symptoms, and look more closely at credit, land tenure, illegal land use, infrastructure, as well as underlying development incentives, among other factors, and then strengthen existing protections afforded forests by protected areas, Indigenous areas, and even multiple use forests.'

Barreto looked at the government's control of hyperinflation in 1994 for examples of how politicians can intervene. 'Generally, you get these changes when there is a social movement,' he said. The Real Plan that ended seemingly uncontrollable hyperinflation was put together by a capable team of politicians. He also suggested that Brazil could learn from its success in eradicating foot and mouth disease, which it did with a national programme launched in 1998, which divided the country into disease-free zones and used vaccines, risk control and prevention measures to progressively demarcate foot and mouth-free zones. By 2018, the World Organisation for Animal Health declared Brazil free of the disease. 'This programme was very intelligent,' Barreto said. 'You could do the same thing with deforestation.'

He also argued that governments, banks and the meat industry should cease buying from municipalities where deforestation is rising or is about to rise. Half of the Amazon's 'deforestation risk' is in just 16 municipalities and the meat industry could operate a much more intelligent prevention system, he said, giving a three-month warning to a municipality to reduce deforestation or have its slaughterhouse temporarily shut down. 'I guarantee that in two months the deforestation would end.'

Barreto also argues in his study that Brazil should make more effective use of an existing but under-enforced tax called the Territorial Rural Property Tax (Imposto Territorial sobre a Propriedade Rural, ITR) to charge big landowners who do not produce much on their land. Enforcing it would, he said, force these landowners to be more productive – thus removing the need to deforest to find more pasture. The tax is federal but can be collected by municipalities and the Brazilian government could have raised R$1–R$1.5 billion in 2017 if the tax had been collected, his research found.

Another element to the issue of cattle ranching is that it is also intrinsically bound up with land grabbing – effectively, the stealing of land, as we saw in Novo Progresso. Land is selected, perhaps in a protected forest or even Indigenous territory, the valuable wood removed by loggers, the rest deforested and burned, and cattle put on it to consolidate ownership. There are Brazilian property laws that date back to the Portuguese kings who once ran the country, and the idea that land has to be productive is deeply embedded in its legal system. And so land grabbers use the rural environmental register, the CAR (Cadastro Ambiental Rural), to claim land they have deforested, using cattle to mark possession as they build up a collection of documents to make it look like they own the land.

For them, it is a good bet. Under the Legal Land Programme, the Lula government (in 2009) and the conservative government of Michel Temer (in 2017), introduced laws making it possible for people who either settled or stole land to buy it from the government cheaply, depending on its size and the year they arrived there. These changes were sold as giving hardworking peasants the right to land they already occupied – and some of those who benefited undoubtedly did so legitimately. But plenty of others were speculators.

The Legal Land Programme names a date before which you need to have occupied the land to get it legalised. But it keeps moving forward. The Lula government moved it to 2004, the Temer administration to 2011. The Bolsonaro government tried to make it 2017. Along with measures to reduce the size of protected reserves, these laws convince land grabbers that land they have illegally occupied will eventually be legalised and soar in value.

Despite what governments have claimed, the Legal Land Programme failed to reduce deforestation – in fact, one study found that Brazil's last change to the law actually increased deforestation. So, on one hand, there is the Territorial Rural Property Tax – a tax 'made to stimulate the more efficient use of land', Barreto said, whose proper implementation the farming lobby has always fought against. And on the other, Brazil lets people steal land. 'It's hard to compete with someone who is getting the land for free,' he pointed out.

Milton Pinheiro, a local man I met at Mauro Lucio Costa's farm, told a story of buying around 150 hectares of land with a long and complex ownership history. At one point, the land had been sold for its wood, squatted and deforested, but someone else claimed ownership and there was now a legal dispute. In the past, Pinheiro said, when the Amazon was divided into 'captaincies' – tracts of land gifted by the Portuguese king to noble families – land owners may have had very little notion of what, in fact, they owned, as they had so much of it. 'The measures were totally random,' he said. 'You'd say, "Your land is as far as you can see."' He described seeing one notary's books, which went back to 1792 and included lists of enslaved people an owner had registered there.

The Amazon has a murky history of property ownership, in which land titles were often fraudulent and notaries paid off to register them, or they were falsified, or referred to land somewhere else – so-called 'flying titles' – or they delineated more land than the owner had actually bought or been given. Pará state prosecutor Jane de Souza told me there is around three times as much area included in land titles in the state as there is actual land.

José Benatti, a professor of environment law at the Federal University of Pará, told me: 'Much of what happened before legitimises what is happening today.' Benatti was president of Pará state's land agency, ITERPA, from 2007 to 2010. He had supported the 2009 Legal Land law but later realised that the pressure for the law to be approved was the result of a rise in Amazon land speculation, following the 2008 financial crash.

He explained how rural properties in the Amazon have all kinds of issues with their ownership documents. The title might not cover the whole area, or could interpose onto other properties, or maybe the proprietor lacks proper land registry documents and just has a sale and purchase agreement. 'The market accepts this, it accepts properties with fragile titles,' Benatti said.

According to one Imazon study, there are at least 22 agencies in charge of land tenure in Brazil. Amazon states can also give out land titles to people who have occupied public land, and until 2019, this was all done on paper. Even digital systems don't always work. In Pará, there are four separate computer systems for land titles that don't talk to each other, all of which government employees need to consult. Even then, they often need to go back to the public records. Nor does Brazil have any kind of central land register.

The afternoon after cycling around the Utinga State Park with Paulo Barreto, I met with his colleague Brenda Brito in the lush, neatly landscaped Residence Park in central Belém – formerly the residence of state governors. She had been a lawyer for Imazon for nearly two decades, specialising in the land issue. Composed, precise and patient, Brito knew the convoluted subject backwards.

In Brazil, she told me, all land belongs to the government unless the person claiming it can prove that at some point in the land's history, the government gave up its possession. 'In principle, it is all public unless you prove the contrary,' she said. She was one of the authors of a study which showed that 21 per cent of the Amazon biome in Brazil is private property, much of which is concentrated around the south and eastern regions. Not coincidentally, these are the most heavily deforested areas.

Cattle Chaos: Corporate Accountability

Being able to prove that the land is being used productively is important, in Brazilian law, to claim possession. 'You have to show that some form of product is being produced there, so cattle becomes the simplest way to do this,' she said. 'This explains why we have such inefficient ranching in the Amazon.' But as the study notes, 'no state prohibits the titling of illegally deforested areas, and most do not require a commitment to recover environmental liabilities before titling.' So it is easy for speculators to steal land, deforest it and put cattle on it, and highly unlikely that officialdom will use this as a reason to prevent them or people they sell to getting eventual title.

Land grabbers and speculators increasingly favour public land that has no official designation – it is neither privately owned, nor a reserve, nor anything else. Brito's study shows these lands make up just 14 per cent of Brazil's Amazon territory, but account for 51 per cent of deforestation. Designating areas of forest as protected reserves, settlement areas or areas where sustainable production can happen involves expense and putting in some kind of infrastructure – one reason why cash-strapped governments have held off. There are 100,000 CARs registered in these areas – that is, farms and ranches that have been added to the rural environmental register in places where it should not be legally possible to do so. Most of these CARs on undesignated land were concentrated in the Amazon's most preserved regions.

Another key issue is uncertainty over ownership. The study found it was not clear who controlled 28.5 per cent of the Amazon – 143 million hectares. People may have bought some of this land in the past and been given titles, but the maps of these properties were not digitised or included in databases – especially those registered before 2002, when geo-referencing was legally required. State governments are in charge of deciding what happens to 17 per cent of the Brazilian Amazon, or 60 per cent of this area, some of which is their property. A Ministry of the Environment study found that 43 per cent of this vast undesigned area – 61 million hectares – should have high priority for conservation. The fact that it doesn't makes it easy pickings for land grabbers.

'What Brazil has to do is designate 60 million hectares of publicly owned forests that at this moment could be targets for land grabbing,' Brito said. The regularisation process takes so long because it's not just simply making a map, checking the CAR and emitting the title,' she said. 'There is a whole job that has to be done to know if that area can actually be legalised.' Brazil needs to change its land ownership and titling laws at the constitutional level, Brito said, so that farmers can only regularise land – that is, secure an official title – if they meet environmental requirements. If they need to reforest or recover a degraded area, the title should be dependent on this being done, with monitoring to ensure the owner loses the land if these requirements are not met.

'All the land laws, not just federal, but of all the Amazon states, need to be altered so that they become compatible with environmental laws,' she said.

The state of Pará is trying to resolve its land ownership mess with an automatic system checking CARs. Since the first half of 2020, some 20,000 registers have been legalised, out of 243,000, according to the state's environment secretary Mauro O'de Almeida. 'Clearly, the more land you legalise, the less environmental problems you will have,' he told me in a video interview while I was in Belém, which O'de Almeida suddenly cut short after about 20 minutes.

Even if farmers are given technical help, credit and a fast track towards regularising their CARs and their land, it's an arduous process. Some will have to pay a fine because of past environmental offences and either don't want to or can't, and others worry they will have to pay a fine if they do legalise their land and get cold feet.

'The other difficulty is that there are people who don't want to legalise, they prefer to work in the disorganisation,' Pará's environment secretary told me, during our short call.

Maurício Fraga Filho wanted an early start, but I wanted to first eat breakfast in my little hotel in Marabá, a brutal concrete city of nearly 300,000 on the banks of the River Tocantins in the south of Pará. And while nothing special in itself – a half-hearted take on the traditional Brazilian hotel spread of fruits, juice, bread, cheese, coffee, cake, chunks of sausage in tomato sauce and scrambled egg – the meal was enlivened by a vividly blue male peacock strutting about on the flat tiled roof just outside the window. The bird opened its wings angrily in a majestic spread of green, grey and yellow. Then the woman preparing breakfast came, opened the window and gave it a piece of cake.

Fraga Filho was outside at the wheel of his shiny white 4x4 pickup, his wife Cecília in the back. I got in and we sped off. Having learnt more about the problems of land ownership and enforcing environmental laws from academics and NGO workers in Bélem, I wanted to understand the pressures on the farmers themselves, and Maurício had agreed to show me his father's farm, a four-hour drive from Marabá. A balding, affable, toothy man in a blue shirt, he had a country burr and a sharp businessman's mind. He was an old friend of Mauro Lúcio Costa – they talked all the time, he told me – and was president of Acripará, a Pará state ranchers' association. I chatted easily to the relaxed and friendly middle-class couple in their fifties while the pickup flew past endless cattle ranches and the odd clump of forest.

After a few hours of smooth tarmac, Fraga Filho began weaving around potholes and ripples in the highway as the road worsened. Then he cut off the highway. 'My father built this road through the forest to reach his farm,' Maurício said, pointing out a towering row of Brazil nut trees. 'This used to be a Brazil nut extraction area.'

His father's farm was 16,000 hectares of rolling Amazon pasture called Fazenda Porangaí, or Porangaí Farm. We pulled up at a low, quietly elegant farmhouse with heavy wooden furniture and sat on a wide terrace that faced a sleek, open-air swimming pool and a row of tall palm trees, home to a semi-tame flock of hyacinth macaws who squawked and chatted furiously throughout our interview.

Maurício grew up in a family of ranchers in São Paulo. His father, also called Maurício, bought this land in 1972, but only began deforesting in 1984. Maurício studied to be a vet and he and Cecília first moved here in 1990, when their firstborn, Gustavo, was a baby. They stayed five years then left. There was no power, light at night came from a generator, they had no television, no telephones. Three times a week, a flight from São Paulo landed in Marabá, bringing a newspaper so the family could catch up on what had happened days earlier. The region was 'practically 100 per cent' forest back then, Maurício said.

Cattle Chaos: Corporate Accountability

His father had progressively cleared all of the forest and now had three farms in the region, including this, the biggest, but never lived here. In 2012, with the children growing up, Maurício and Cecília moved back to Marabá and had been there ever since. Gustavo now managed his grandfather's ranch but Maurício was very much in charge. He also owned half of another ranch further south in the state, which his partner and brother-in-law ran.

Porangaí's business was based on buying young cattle cheap, that had already been fattened up to a certain point, rather than birthing them. Maurício said he knew the farmers the farm was buying from – the indirect suppliers – but had no idea who they had bought from. I asked if he could have bought cattle raised illegally inside a protected area or even an Indigenous territory. 'It's possible because I don't know,' Maurício said. 'We were never worried about this.' He said the farm buys cattle from other supplier farms or smallholders that have almost certainly deforested illegally, generating an alert on satellite imagery. Locals call this having a 'PRODES', after the satellite system that provides Brazil's most accurate annual deforestation numbers. 'An infinite number have this. All of them must have PRODES. Nobody will have any documents. Nobody is able to legalise themselves.' If JBS and the other meat companies started properly monitoring their full supply chains, he would be unable to sell to them. 'On the day they monitor the indirect suppliers properly, I am out of the market, me and everyone else,' he said.

The big meat companies in 2009 promised not to buy cattle from farms with any illegal deforestation, or that had been accused of employing workers in conditions 'similar to slavery'. But it is not a legal requirement, Maurício noted. I told him JBS and other big companies had promised to be monitoring all their indirect suppliers by 2025. 'They won't resolve this,' he said. 'There is no way. They can't.'

Earlier the same year, a new scheme had been launched in Pará on a test basis called Selo Verde – Green Seal. It had been set up by researchers from the Federal University of Minas Gerais using public records. Anyone with the number of a CAR (rural environmental register, which farmers register online themselves) of a property could check if a farm had any fines or embargoes. They could also see if it had sufficient legal reserves – the widely-flouted requirement that a certain proportion of the land be left as forest – and even check if it was sourcing cattle from indirect suppliers that could be reputationally contaminated with deforestation. Maurício loaded his farm up on his cellphone, and the Selo Verde app showed that his farm had problems. Maurício and other powerful ranchers were dead against the system and had met the state governor, Helder Barbalho, to voice their objections.

Under Brazil's Forest Code, Fazenda Porangaí should have had a 'legal reserve' of natural forest or vegetation of 50 per cent or more, but native forest made up just 29 per cent of its area. Selo Verde flagged this. Also, Selo Verde said more than 20 per cent of the cattle it bought came from indirect suppliers and were, as Maurício himself said, likely to be contaminated by deforestation.

According to Selo Verde, the 8,000-hectare ranch he owns with his brother-in-law near Redenção has an environmental deficit of nearly 2,000 hectares in its legal reserves and the areas farmers are required to leave as forest around streams, slopes and hills, called permanent protection areas (*áreas de proteção permanente*, APPs). One of his father's other farms, the 7,000-hectare

Rita de Cássia, was flagged on Selo Verde because more than 20 per cent of its cattle came from indirect suppliers and were therefore potentially contaminated by illegal deforestation. His father's other farm, Sinhá Moça, also has an environmental deficit and its CAR is 'pending'.

The state government of Pará is in the midst of the gargantuan task of regularising all the CARs for each farm and checking them against deeds – and in this process, it will eventually reach Porangaí. At that point, its owners will be invited to join the Environmental Regularisation Programme (Programa de Regularização Ambiental, PRA) and find a way to increase its legal reserve, possibly by reforesting or negotiating other environmental commitments.

'Anyone who has an environmental liability only gets rid of it when they legalise and adhere to the PRA,' Maurício said. He said if Selo Verde or any other scheme – such as those the meat companies say they are introducing – to control the indirect suppliers is introduced, large numbers of farmers would be flung into illegality, at which point they would simply sell their meat on the clandestine market. Maurício argued – and his argument is widely used by Amazon farmers – that before cattle ranchers in the most threatened and biodiverse tropical forest in the world can actually be sure they are not buying cattle from illegally deforested land, local government has to finish regularising nearly a quarter of a million self-registered environmental registers, many of which were simply plonked on top of protected reserves, or are on land where other titles already exist.

That process could take a decade or more. And the institutional and legalistic mess that Amazon property ownership is in is, in turn, the result of 500 years of state-sponsored migration, colonialism, chaos, land grabbing, fraud, chaos and confusion over land titles, not to mention the violent expulsion of some communities, settlers or Indigenous peoples. That sounded like a reason to simply not do anything, ever, about a problem Amazon cattle ranchers simply don't see as anything to do with them. Maurício didn't see it like that.

He thought it would be possible in theory to trace where cattle they were buying came from as long as the seller had not bought from a sort of cattle wholesaler or distributor, common in Pará, called a *catireiro*. The *catireiro* assembles lots of cattle bought from smaller distributors, who have driven a truck collecting the animals from smallholders and little farms. 'He is a cattle seller,' Maurício said. 'He doesn't know where they came from. He doesn't know and neither do I.'

Later, Maurício introduced me to a *catireiro* in Marabáthat his farm buys from, called José Hilário Andrade. 'I buy from the smaller guys,' José told me. 'I buy three from one, sometimes ten from another, twenty.' He puts a lot together of anything from 50 to 200 head of cattle and offers them to farmers like Maurício. He had around 20 customers like him – 'big farmers,' he said – and sourced from around 300 smallholders, selling 4,000–5,000 head of cattle a year. I asked about the possibilities of tracing these cattle. 'I don't think it would be easy,' José said. 'You'd have to formalise a lot of things but I think it could be done.'

José said he had no idea if the smallholders he bought from had environmental problems or not because that was the meatpackers' job to verify. He insisted all the cattle were properly vaccinated. José later sent me examples of three GTAs – one with

Cattle Chaos: Corporate Accountability

10 cattle on it, one with 60, and another with 80. The GTAs were all in the name of his wife Marcilene Santos – she had a small farm, he explained.

'I buy ten from one, five from another . . . and they are passed onto my records because I have records, a small farm, right? . . . I put them together with 100 animals from my property, for example, and I sell to Maurício,' he said. 'That GTA on my records, which I drew up for my property, I am going to transfer now to Maurício's records.' It did not sound like there was very much control at all over the cattle movements around Pará.

Back at Porangaí, it was lunchtime. Places had been laid on a counter in a shaded area near the pool, where the rest of the family was waiting. Two young Black women in pink maids' uniforms served tender, delicious slivers of barbecued steak. The social dynamic was archaic: the Black servants serving the rich white family. I congratulated Maurício on the tenderness of his meat. 'It's not from here,' he said. This was a more expensive variety of beef, called 'prime', sourced from another ranch. The family didn't eat their own meat. They just sold it to others.

After lunch, the conversation turned to global warming and the argument made by climate scientists like Carlos Nobre that the Amazon could be within a decade or so of its 'tipping point', after which it starts degenerating. Maurício didn't buy it. He was not convinced that climate change had a human cause and did not seem remotely worried about it. Brazilian climate change deniers often argue that planet Earth has gone through cyclical changes before, and that global warming is just the latest of these. Maurício had clearly encountered this argument.

'I've listened to Carlos Nobre a lot and I don't know,' Maurício said. 'The climate changes. Brazil or the world has gone through ice ages . . . is this all really the Amazon?'

However, Maurício was worried about the impact on business. He had met recently with Brazil's then agriculture minister, Tereza Cristina Dias, and a senior official and heard their concerns that, for instance, China might stop buying Brazilian beef or that the state of Pará might be banned from exporting beef. He wondered if farmers outside of the Amazon might start to disassociate from rainforest ranchers and treat them like 'lepers'.

'I am very concerned,' Maurício said. 'But I think the environmental regularisation has to come at the same time.' He argued that meeting the standard to legalise a farm should be easier, that standards should be loosened so they could be met. There was no point having 'tough legislation that nobody manages to obey'.

Instead, he wanted to show me how, like his friend Mauro Lúcio Costa, he was improving the pasture on the farm so as to intensify production. The family did not want any more farms, they wanted more meat from the farms they had. This would be their contribution to the deforestation crisis. 'We have to intensify, we believe in this,' he said.

So we headed off in his 4x4, bouncing down bumpy dirt roads deep into his farm. It did not look like the Amazon. Instead, with its rolling fields, dotted with trees and spotted with white cows, Porangaí Farm could have been Europe or the United States, were it not for the palm trees and the wildlife. Maurício pointed out a caiman lurking in a pond and a tapir bathing in small lagoon. We passed an area of pasture that had become overgrown, reaching twice the height of a man's head with trees as thick as a leg. Along

Cattle Chaos: Corporate Accountability

the way, Maurício proudly pointed out the maintained forests and wetlands around streams

Some fields had been ploughed by tractors with enormous wheels that threw up clouds of red dust. Smaller tractors with GPS devices were spraying seeds in another field. On a nearby hilltop, a herd of peccaries, a breed called *caititu* here, stood watchfully. Perhaps they were wondering where they would forage now. In between driving from field to field, meeting workers and watching tractors, Maurício explained the two intensification techniques available – he could either fertilise the soil and plant grasses developed in Africa and Australia, or confine the cattle and give them feed.

Down at the scruffiest end of the farm, a rough, hilly field was being cleared when we pulled up. Two small diggers with caterpillar tracks were pulling an enormous chain between them – a device used by loggers to clear forest. The chain ripped up undergrowth, bushes and trees that were three to four metres high while engines screamed. An orange-beaked, white-necked carrion hawk, called a *carcará* in Brazil, circled above the newly cleared land, aware a whole bunch of small animals had suddenly lost their homes and were now easy meat.

The brutal, mechanised clearance of this field stayed with me. Later, in a phone call, I asked Maurício if the family had considered leaving it to grow – with a decade or so, those skinny trees would have become a new forest, and he would be well on the way to resolving his environmental liability.

'No, this option is economically very bad,' he said. 'I have all the structure there, there is a corral, there are fences.' Instead, the family will wait until the state of Pará gets around to validating their CAR and tells them what they need to do as part of the environmental regularising programme. 'Then we'll decide if we want to compensate, if we are going to let [an area] regenerate.'

His plan to resolve the 'environmental deficits' on the family's farms was by buying land inside a reserve which he would leave untouched. He had bought 566 hectares of farm inside a reserve in the nearby Amazonas state that he could count towards his legal reserve. But this farm alone had, at the most optimistic calculation, a deficit of legal reserve of nearly 3,500 hectares.

It seemed like Porangaí and the other farms were going to carry on just as they were while the government resolved the land mess. This was, after all, a business. But even efficient businesses like Fazenda Porangaí can fall foul of the Brazilian authorities. In June 2018, while his father, Maurício Pompéia Fraga, was on a cruise ship in Italy, government inspectors found what they said were 30 of Porangaí's employees, including a 16-year-old boy, working in what Brazilian law calls 'slavery-like conditions'. They were riding horses and driving cattle 900km from Uruará, far away in Pará, to the ranch. Maurício had told me the farm sourced cattle from the area but neglected to mention how they got there.

During the four-month journey, labour inspectors said the workers slept under tarpaulins, with no running water or toilets, eating food provided from a cart that rode alongside. For this they were to be paid a day rate of 45–60 reais (US$12–16) on arrival. In 2021, the farm and Maurício Snr were placed on the government's annual 'dirty list' of people, farms or companies who had subjected people working for them to 'slavery-like' conditions.

JBS and Marfrig blocked the farm as a supplier. It could then only sell cattle to other ranchers. 'It's a very serious problem,'

Maurício said. He argued that cattle drives like this – called a *boiada* – had historically been a tradition in the region. 'It was a normal practice that everybody knew about. Then came a labour inspector who thought it was slave work,' he said. The farm had not been legally condemned – in Brazil, where court cases can take decades, the decision to put someone on the list happens first and they fight the decision in court. A subcontractor had hired the men, Maurício said, they did not work directly for the farm. No employee had ever taken them to court.

He seemed as unconcerned about the workers as he was about global warming. 'The employer will always look for someone who will work for a lower price, a lower value, just as when we go to the supermarket, or a shop, or to contract a service, we always look for the cheapest,' he said. 'That's the way it works. If there is someone who will do it for 40, I won't pay 60.'

That is the thinking that has got Amazon cattle ranching into this mess in the first place: money above all. If Maurício didn't really believe in the climate emergency, he wouldn't do anything to stop his farms contributing to it. If he could avoid letting old pasture grow into new forest, he would do so, because he saw that as a waste of land that could be making money. He didn't see the long-term cost to him and his family. He hadn't priced climate change into his business plans. By the time he does, it may be too late.

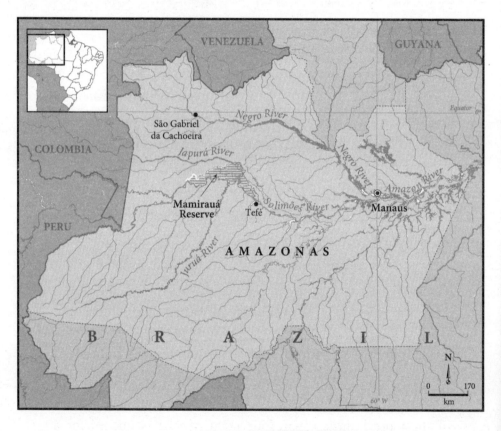

Amazônas State

CHAPTER THREE

Putting the Eco Back in Economy: Agroforestry Models

Dom Phillips
Marabá, Pará

'The best way to convince a farmer is with a farmer.'

When I first came to the Amazon, I was dumbfounded to find that it had cities as big as Marabá. Until recently, though, things were very different. For centuries, this region in the state of Pará was home to diverse Indigenous groups and their forest and savannah homelands. Marabá only became Marabá in 1913 and, even then, for decades it remained a frontier town, only growing as Indigenous peoples like the Gavião were decimated by illness, oppression, murder, ranching, logging, land invasions and forced assimilation. Now, 266,533 people live in the city, and there isn't much rainforest left. Instead, it's a city that hums with traffic, split in half by a busy highway, with cellphone antennas looming over houses with red-tiled roofs.

In a pretty little square one night, teenagers queued for a booming, neon-lit disco bus. Families crowded a riverside fish restaurant. Bars served beef and played country music. This was a city with a small-town mentality, I thought, as I washed clothes in an automatic launderette in the basement of an overly air-conditioned department store where hip hop videos blasted from a big screen but the cellphone signal froze. If this was progress, something had gone awry.

Marabá's branch of Brazil's Havan department store chain had its trademark Roman arches and 37-metre-high replica Statue of Liberty, just like Havan stores in other Amazon towns. The chain's owner, Luciano Hang, was a high-profile figure, a bald, rabid supporter of far-right president Jair Bolsonaro, who wore the loud uniform of a comic book character, a suit in patriotic yellow and green, and paraded equally extreme views on social media. For shoppers, it may have been just another store, but to me Hang and Havan always seemed emblematic not just of how powerful Bolsonaro and his destructive thinking were in this part of the world, but of how many people preferred their infantile model of the future, based on relentless development at any cost, more guns and a notion of freedom as fake as that Statue of Liberty.

But Marabá was still surrounded by countryside and I was there to examine the workings of the bio-economy, the use of renewable biological resources to promote a sustainable, circular economy. This concept encompasses everything from biotechnology to community cooperatives. The meaning seems to change depending on who you ask for a definition, but despite this vagueness, there is widespread acceptance that it is important. And if the Amazonian bioeconomy is – as many hope – to have a chance of helping save what's left of the rainforest, it has to work somewhere like this. Especially somewhere like this. That meant a few more days in Marabá. And so, one Tuesday morning, I sped down the highway in a 4x4 driven by Daniel Mangas, a phlegmatic, grey-bearded technician of 59 from the Brazilian agricultural research institute Embrapa, in search of a different future. We passed cattle pasture, the odd clump of trees and *babaçu*, a kind of palm. Eventually, Mangas turned off down a dirt road, past a grubby motel called Aphrodite offering rooms by the hour.

The Mamuí Settlement Project consisted of several thousand hectares of rolling hills, cattle pasture and little patches of forest cut by dirt tracks. It was the winter dry season, and it was dry. Antonio Mauricio Batista let the way down the parched, grassy bank into the little valley, a scythe in his hand. A lithe, enthusiastic man in his late forties in a baseball cap and a light red jacket, he wanted to show me something hugely important to him on the 26-hectare plot of land that he and his family farm, an hour from Marabá. It was just after ten in the morning, the sun was high and before us was a hill of scrub and grasses in different shades of green and yellow. Apart from a smudge of dark green on a distant skyline and a row of palms on the hilltop, there was little to suggest that we were still in the Amazon.

But at the bottom of the valley was a shady copse, including the super-fruit palm açaí, the *buriti* palm, cacao (cocoa, *cacau* in Portuguese) fruit trees, and native Amazon *andiroba* and *jatobá* trees, all growing around a clear spring bubbling fresh water. The air was cool and fresh down here, where a blue dragonfly hovered elegantly over the water. Instead of dry earth, the ground was damp and muddy because Mauricio – as he was known – had made a dried-out spring flow again.

'The water came back; it came back to life because of this recuperation,' Mauricio said as he poked in the water with his scythe. 'That's what we have to do, protect our nature, our springs, so we don't lack water for our children, our grandchildren in the future. Isn't that right?'

Mauricio's previous attempts to plant *buriti* palms in the dry valley floor had come to nothing, but then, in 2019, a team from Embrapa, the Brazilian agricultural research institute, advised him to seal off the valley from his cattle and plant native trees. So he did. And the water began to flow again. 'Look at the açaí tree there. It was just thrown there and it was born,' Mauricio said, pointing out the palm that produces the 'superfood' berry eaten across Brazil. 'Natural regeneration!' he exclaimed, then stepped on a rotten log which crumpled under his weight so that his boots sloshed into the water. We chuckled. 'If you don't get wet, there's no water,' Mauricio smiled.

Back above the valley, Mauricio showed me how the new spring was feeding a small reservoir which, during this dry season, irrigated an agroforestry plot he was growing. Agroforestry means farming trees, shrubs and crops together, in a man-made recreation of a natural forest that is designed to produce. It is an ancient system increasingly being revived and widely presented as a way to restore degraded pasture – overgrown, abandoned cattle land – of which the Amazon has plenty. Mauricio's new plot was coming on well. He proudly showed us the banana trees weighed with big bunches of fruit he was already selling and the tiny, perfect cacao plant buds on a tree already a couple of metres high, whose seeds make chocolate. Beans, corn, pumpkin, okra twisting on a vine, *jiló* – a bell pepper-shaped green vegetable with a refreshing taste – and *maxixe*, a spiny green ball not unlike a cucumber, grew lower down. The vegetables, Mauricio said, were 'to eat while the fruit grows'.

Putting the Eco Back in Economy: Agroforestry Models

At his house, we sat at a rough-hewn wooden table under a mango tree in the yard and his wife, Glenilda, served chilled, tangy *acerola* juice. The couple had four grown-up children. Much like the settlement project Esperança I visited in Anapu, Mamuí was created for landless peoples, the rural poor, by the national agrarian reform programme. Mauricio and others had occupied the land in 2003. It was a cattle ranch that had fallen into disuse, and they had heard the government land agency INCRA would take it over and divide it up for people like them. They decided to get in there first – before the authorities had finished the process, a tactic that eventually paid off, but had involved a good deal of suffering.

'Life was really tough because when we came here, we had no house, no infrastructure, and we worked. Thank God that today we live from the property and I don't need to go anywhere and work for anyone else,' Mauricio said. In 2017, the settlers here were given title to the land. Each now had a plot of around 25 hectares and most farmed cattle.

The Amazon is made up of different ecosystems with different climates. This region is a 'transition zone' between rainforest and the *cerrado*, or savannah, just south of here, which meant the trees weren't as tall as thicker forest in the Amazon's centre, the vegetation was a little less dense and the region had a distinct dry season.

When Mauricio and his companions first occupied the area, some of the land was degraded, overgrown cattle pasture and some was still forest. Within a few years, they had cleared most of it. On a nearby farm, I met João Pinheiro, a community leader everyone simply called 'Sadia', a gaunt, self-possessed and thoughtful man in his sixties. He told me why this happened: they felt they had no choice. It is typical of Brazil's often schizophrenic policies towards the Amazon that not that long ago, the government land agency INCRA encouraged people like Sadia to deforest to prove they were using the land they had been given – or lose it.

'We pressured everyone to deforest,' Sadia said, 'to show INCRA that we wanted this to survive.' Banks also encouraged settlers like him to deforest and raise cattle in order to get financing, he said. He felt differently about this now. 'Everything that God created over millions of years we destroyed in a few years,' Sadia admitted. 'Deforestation is the greed of man and a lack of guidance by INCRA and the bank which finances farmers.'

Now Sadia was growing açaí palms along a watercourse and fertilising them with manure from his small herd of cows. Sadia had less native forest on his plot than the 80 per cent required by Brazilian law (50 per cent if the deforesting was done before 1996), but the açaí trees would help reduce the deficit, as well as bringing in income. 'They advised us that we have to reforest to be within the law and it just needed a little push for us to do it,' he said. 'Today for us this is a dream come true.'

As an outsider, it can be easy to react in horror to any and all destruction of the Amazon. But I felt it important to understand more about the lives of people like Sadia, the grim poverty that drove them and their families to migrate to the Amazon, and the way the state and society encouraged them to deforest everything. Again, like Tunica, who I met at Esperança, Sadia had grown up in the northeastern state of Maranhão, the poorest state in Brazil. Sadia left at 18 and came to Pará in search of a better standard of living, on a frontier with freedom and opportunity, he said. He worked on farms and as a *garimpeiro* (wildcat gold miner), hand-digging jungle mines. 'It was the height of gold. Everything at that time was gold,' he said.

Sadia worked at Serra Pelada, a notorious open-cast gold mine where tens of thousands of men teemed in at the bottom of a deep, muddy pit. The gold rush there had begun when a child swimming found a six-gram nugget of gold. Swarms of people flooded to the region. One nugget found there weighed nearly seven kilos and the mine became so overrun and chaotic that it was put under military control during the dying days of Brazil's dictatorship – which ended in 1985. Even so, deadly accidents were occupational hazards and Sadia saw men buried alive more than once. In 1986, the great Brazilian photographer Sebastião Salgado captured hellish images which made Serra Pelada world famous. His epic photographs of hordes of men toiling like ants in the mud and ascending wooden ladders carrying sacks of rocks and earth on their backs were later compared to the paintings of Hieronymus Bosch.

Sitting outside the house he shared with his wife Maria Pinheiro, Sadia was philosophical about his gold rush experiences. 'I made a lot of money. But we don't know if it's gold money that is crazy or us that are crazy, we end up spending it all in pain, throwing it all away,' he said. 'When the gold ended, things got difficult and every *garimpeiro* was screwed. Then most of them put their feet on the ground and became small farmers, just like me.'

Now, Sadia and many others said they want to change what they do. Climate change was becoming very real to them and they felt it as a threat. 'Summer got longer, in the past it didn't last as long,' said Sadia. The rains come later and they are heavier. 'Man is destroying the planet, all for greed and money.'

Their kitchen was in an outhouse made of bricks and concrete, left half open to the elements. Earlier, Maria had made lunch, cooking on a hotplate over the log oven typical of these parts – fish in batter, traditional beans in a meaty sauce, rice and salad. There was a fiery homemade chilli sauce on the table and we drank fresh juice. She was warm and chatty, full of stories and gossip about the neighbours. No, she would not accept any money for the lunch. Instead, she pointed to a hammock and said if I wanted a little siesta I should feel free to lie down in it. Maria showed us her allotment, with its neat rows of fruit and vegetables under nets to keep birds away. She used to sell at the street market in Itupiranga, the nearest town, but now she was selling to the state government for school lunches. 'It's great because what we produce was going unused,' she said, showing us invoices of crops she had grown and sold – 26 kilos of *galego* limes, three kilos of *maxixe*, 95 kilos of green coconuts. Feeding schoolchildren with healthy produce from family farms and kitchen gardens is a successful policy that Brazil has adopted in rural areas. For Maria, it had also transformed her income.

Sadia and Maria wouldn't let us leave. They kept offering more – coffee, coconut water, fruit – to get Daniel and me to stay longer, each suggestion a fresh gambit. One of them would point at us and say: 'juice', and motion us both to a chair, as if the argument was settled. Their hospitality was as warm as Maria Pinheiro's smile and her homemade chilli sauce.

The Embrapa project helping Mamuí is called InovaFlora and is part of a wider initiative, called the Integrated Project for Sustainable Production and Management in the Amazon Biome – a clumsy

Putting the Eco Back in Economy: Agroforestry Models

and intimidating title typical of Brazilian government-speak. But despite its relatively small scale and budget – just R$8.6 million – it was clearly proving successful in Mamuí. The problem was that funding came from the Amazon Fund, an international scheme under which Germany and Norway gave money to Brazil for reducing deforestation. During the Bolsonaro administration, the environment minister Ricardo Salles effectively froze the scheme for ideological reasons. Salles claimed he had found irregularities in spending, but never produced any evidence; Bolsonaro had for years railed against foreign powers interfering in the Amazon, also without ever providing any evidence. InovaFlora continued where other projects had withered.

In a small group of houses around a baked earth yard, Elias Barros da Silva picked up a large machete, put on a wide-brimmed straw hat, and led the way down the hill to show us his protected area – a big, fenced-off area near a pond full of high, skinny trees and luxuriant foliage. It was once old pasture overgrown with the rough undergrowth people here called *junqueira*. First, Elias sowed more grass for cattle. Then, when the Embrapa project started, he planted trees. Now it was beginning to look like a forest again and Elias, with his shy smile and gentle demeanour, was quietly proud of it.

'We were being encouraged to plant and I said, "I'm going to plant,"' he said, listing açaí and *cupuaçu* – a rich, creamy fruit – and trees including *andiroba*, mahogany, *amarelão*, and *jatobá*, while fingering the blade of his machete. 'There are many species here,' he said. Beside it was his agroforestry patch, a small plantation full of fruit trees. There were banana trees and he was already selling their fruit, cacao was growing healthily and there was sugar cane.

And now there was water too. 'There is a spring there, just there,' he said.

Elias sat on a log in the late afternoon sun. Now in his late thirties, he was the youngest of seven children and was a toddler when his father Raimundo abandoned his young family in Maranhão and his mother, Antônia, headed for Pará because she thought the family would starve otherwise. Elias had never met his father. For three years, Elias worked producing charcoal, formed in intense heat in round earthen mounds. Not that far from Marabá there was a highly pollutant pig iron industry that depended on the fuel. 'Charcoal at the time was a product that brought in a lot of income,' Elias said. He carried on producing charcoal after he bought 15 hectares in the settlement from his father-in-law, even though, in effect, this meant he and others were burning what was left of their own forest. 'There were always things being done that were wrong,' he said.

In 2009, under the Action Plan for the Prevention and Control of Deforestation, a major police operation called Arc of Fire targeted the region. Elias and his charcoal colleagues decided to cut their losses. 'We decided to stop and just invest in agriculture and raising cattle,' he said. Now he was selling his bananas from quality seedlings provided by Embrapa and he wanted to keep the cattle. 'It's a very secure source of income,' he said. 'I'm selling cheese and I'm selling milk.' We finished the afternoon on the shady terrace of his simple family breezeblock house, overlooking the earthen yard. Elias served coffee and slices of his homemade cheese, rich, salty and delicious.

Sitting there on the veranda with the cheese and hot, sweet coffee, listening to Elias talk about his fruit trees, it felt like the

Putting the Eco Back in Economy: Agroforestry Models

wheel was beginning to turn, that those who cleared the forest for the reasons described to me by farmers were willing to start reversing that process – provided they could get help to do so and still produce enough to live on. Elias had 25 hectares in total, 15 of which were titled by INCRA. Publicly available land records on those 15 hectares confirmed what Elias had told me, that nearly six hectares – 37 per cent – had either been left as a clump of forest or were growing with his agroforestry system. (The other ten he bought bit by bit were cleared pasture or overgrown pasture and are still registered in the name of the neighbour who sold them.) It looked to me like a little success story, a little ray of hope glinting off the açaí palms in a region that desperately needs it.

There is a lot of talk around agroforestry systems as one of the solutions to the Amazon's particular matrix of problems – the relentless deforestation and the grinding poverty. It's easy to see why. Agroforestry systems give Amazon farmers the chance to recover cleared land, restore the rainforest and make money at the same time. But I felt it was important to understand more about what agroforestry actually is and what benefits it could offer – as well as those it can't. So I turned to Joice Ferreira, a biologist, ecologist and researcher at Embrapa in Belém, a warm and patient woman with formidable Amazon experience and knowledge who has studied Amazon agroforestry projects extensively.

It is important to understand, she said when we met in a park in Belém, that agroforestry means creating a plantation – not a new forest. 'Agroforestry systems fall very well into the category of restoring farming environments. They have enormous advantages for farming environments. But assuming they mean restoration without questioning what the limits of this are is very risky,' she said. 'Even if it is really well done [an agroforestry system] is hard to compare to a forest.'

Joice grew up in the neighbouring state of Tocantins but lived in Belém. She was a member of the Science Panel for the Amazon, an international group of leading scientists who wrote an extensive Amazon Assessment Report for the 2021 COP26 climate conference – Ferreira co-wrote the chapter on bioeconomy. She was also a member of the Amazon Concertation, a network of academics and other individuals, organisations and companies studying Amazon solutions. If the Amazon is to be saved, scientists like Ferreira who know and understand the rainforest, its people and its ecology have much to contribute.

'The transition to more sustainable forms of crop and livestock production – i.e. that conserve the natural environment, minimise climate change, reduce impacts on biodiversity and benefit Indigenous people and local communities – must be at the heart of Brazilian concerns. Agroforestry systems, if well planned and managed, score in all these aspects: they increase biodiversity, carbon stocks and have great potential to restore degraded agricultural areas,' Ferreira and two other researchers wrote in an article in April 2022.

None of this matters, though, Ferreira said, if Brazil can't get deforestation under control. 'Everyone says bioeconomy is the solution for the Amazon,' she said. 'But nothing by itself is going to save it if the problems continue.' Studies have shown that restoring forests makes no difference when they are being razed at the rate they currently are. 'It only makes sense to talk about restoration when deforestation is controlled.'

Putting the Eco Back in Economy: Agroforestry Models

But agroforestry systems do have many advantages when used to restore land degraded by single-use or monoculture farming, or by cattle ranching, like I saw at Mamuí. Studies back this up. One showed that agroforestry systems that included oil palm were more efficient at storing carbon than even secondary – or re-grown – forest. Another showed how farmers raising fruit and black pepper had a per-hectare income nearly four times higher than soya farmers and nearly nine times that of cattle ranchers.

While agroforestry should never be used to replace existing forest, as Ferreira explained, together with measures discussed in other chapters – like reducing deforestation and increasing control of the cattle supply chain – it can help protect the Amazon. Agroforestry helps with water storage, too, research in Indonesia and Uganda has found, because it helps protect water sources and its shade protects moisture in soil. And it can provide income for smallholders – as I saw at Mamuí, where within less than two years of starting planting, the farmers were already able to sell bananas.

Life was tough for Amazon smallholders. Ferreira said that when asked how things are going, they often quip 'I'm escaping' – an example of how close to the breadline they live. According to Brazilian government figures from 2017, the nine states that make up the Brazilian Amazon had an average human development index of 0.750, compared to Brazil's average of 0.778. Pará was the fourth worst-off state of all 27 in Brazil. So convincing some of Pará's poorest residents to make a risky switch to agroforestry, or to recover degraded pasture with agroforestry, meant not only convincing farmers raised on the dream of pasture for cattle, but giving them advice and technical help over a long enough period to make sure the change worked for them.

For example, one issue is that when fruit trees grow higher, the amount of shade beneath them increases, and the vegetables grown beneath have to find a home elsewhere. Farmers need to understand this, Ferreira told me. She co-wrote a 2019 report on the successes and failures of three agroforestry-based restoration projects in the Marabá region. In the case of one, forest cover decreased by half: more income does not necessarily mean less deforestation. Some agroforestry projects for smallholders she studied floundered because they only lasted a couple of years, as people lost interest or reverted to cattle.

'Any project that involves trees has to be long-term, you have to accompany it, you have to monitor it to know what worked, what didn't, what could be done,' said Ferreira.

These provisos aside, there is vast potential for agroforestry in the Amazon. A study on Mamuí projected that within three years, its farmers could be producing 100 kilos of cacao per hectare – and 500 kilos within seven years. Açaí production was also expected to grow – with farmers expected to produce 1,800 kilos a hectare in four years and 5,000 kilos in nine years. It is to be hoped Embrapa follow the farmers long enough for that to happen. As the Science Panel for the Amazon report Ferreira co-wrote said: 'The greatest chances to reposition Latin America from a commodity-based economy toward a nature-based economy is through the conservation of its natural resources and, above all, the application of science and technology.'

Açaí alone is already big business. Apart from sales to successful exporters like the US company Sambazon, there are açaí stalls all over Brazil and the superfruit is popular with sportspeople. Studies have suggested it may even slow the development of prostate cancer and its seeds were being tested for everything from making furniture to bricks. There are issues around children used in some regions to harvest the fruit and there are massive açaí monoculture plantations, but there is still a lot of hope around the fruit's potential to help smallholders. In 2019, the gross income from açaí production in the Amazon region was R$3.02 billion – US$765 million – and the global market could reach US$2.1 billion by 2025. 'Everybody wants to plant açaí, it has this flagship appeal. So does *cacau*, there always has to be a more valuable plant,' said Ferreira.

Motorbike taxis were a cheap, fast and fun way to get around Brazilian towns and cities, but required steely nerves. As a passenger, I liked to be sociable but fretted when the friendly driver I was chatting with took one arm off their handlebars to gesticulate with. When I lived in Rio de Janeiro, I used the sweaty communal helmet the local *mototaxi* firm left on a traffic cone until Covid came along. Here in the Amazon, everyone used motorbikes and *mototaxis* were everywhere, but helmets were harder to find; some regular passengers bought their own.

I was in a village called Quatro Bocas – Four Mouths – near Tomé-Açu, a farming municipality a four-hour bus ride and one ferry crossing from Belém, the Pará state capital. I was here because this region is famous for its agroforestry production, and a cooperative of Japanese immigrant farms responsible for it. Michinori Konagano's farm was out of town and so my hotel's receptionist suggested a motorbike taxi. The driver had a yellow uniform but no helmet for me, so I jumped on, hoping it was down a quiet road.

I relaxed as we eased through the small-town traffic, but once we were barrelling down the main highway, the wind ripping through my hair, and the speedo wobbling around 100km/h, I wasn't so calm. Then the driver turned onto a dirt track and took a call on his cellphone as we bounded down a hill, using one hand to hold the device to his ear and the other to steer, while I sat behind him wincing at every bounce. It was a relief when we reached the low, white house with a red-tiled roof and a wide, open terrace, surrounded by farm buildings, where Michinori Konagano lived.

Michinori was one of the most celebrated agroforestry farmers in Tomé-Açu. He had been president of its agricultural cooperative of producers and agriculture secretary for Tomé-Açu, and he had lectured extensively on agroforestry production. He was waiting on his porch, a thin, quietly spoken and serious man in his early sixties with a natural authority about him. Speaking quietly with a heavy Japanese accent, Michinori came across as someone who never needed to raise his voice. He had a lot to say and something about the way he said it inspired patient listening.

He arrived in Brazil aged two, when his parents moved from Kagoshima in southern Japan to Pará. A Japanese community had been established in what became Tomé-Açu in 1929, when 189 people – 43 families – came to build new lives in the Amazon rainforest following a request from state governor Antônio Emiliano de Sousa to the Japanese ambassador Shichita Tatsuke. The Mixed Agricultural Cooperative of Tomé-Açu (Cooperativa Agrícola Mista de Tomé-Açu, CAMTA) was founded two years later, and the region now had Brazil's third-largest Japanese community. There was even a Buddhist temple in the town.

Putting the Eco Back in Economy: Agroforestry Models

The family was poor and Michiori remembered 'nearly 20 years of suffering'. His father drank *cachaça* – Brazilian sugar cane rum. His mother, Mitico, taught them to be respectful, hardworking and obedient. 'It's humiliating to be poor,' said Michinori, remembering how he fished in local rivers and caught birds in traps to help feed the family while his father worked on a farm. Michinori was recovering from Covid-19 and still looked a little frail when we met, his kidney was badly affected and he had lost 11 kilos. A month before we met, his mother had died of Covid-19 at 84 years old.

Michinori told how he left home at 17 and rented some land from a neighbour. He bought his first land at 19, borrowing money from the Japanese government to farm three hectares of passion fruit. He also grew cacao, papaya, pumpkin and melon. He paid off bank loans. He believed in hard work and dedication, and helping people better themselves – here he reached for a widely used Brazilian saying about teaching people to fish rather than giving them free fish. Life had been hard for the family as for a long time they had specialised, as many of the Japanese here had done for decades, on growing black pepper. But black pepper plants were killed by a fungus and it wasn't until the community began varying the crops it grew in an agroforestry system in the 1980s that life got easier. 'We started to diversify our property,' Michinori said. 'We improved our financial situation.'

Michinori was meticulous about his farming. He pulled out handwritten rain charts to show me how climate change affected the weather here – rainfall for July in the dry season, for instance, varied wildly over the last five years. 'Nobody knows if it's winter or summer,' he told me. 'This is the result of man's action. Man really messes with nature.' It was August and would normally be the dry season in this region, but Michinori could smell rain on the wind.

I met his wife, Amélia, and two daughters in their twenties, Noemy and Mayumi, over a lunch that combined traditional Brazilian and Japanese food. There was fried, battered fish, sticky rice, aubergine salad, black beans and cold, fresh, vivid purple açaí drunk as a thick juice sweetened by honey produced on the farm. It was so good I had second helpings. Michinori and Amélia said little, but their daughters chatted easily.

After lunch, Michinori put on a straw hat and drove me around the 230 hectares he farmed using agroforestry, out of the total 850 hectares he owns – he said around half of the property has been left as a reserve. He got out of the car to show me how trees and plants grew together in rows in his neatly ordered plantation. It didn't look much like a forest to me, but nor did it look like a traditional farm. It felt more like a garden. And it was a lot cooler than the flat fields around it. 'Here we have various cultures planted at the same time in the same area. Black pepper, passion fruit and watermelon were planted here, pumpkin, today we have açaí, *cacau*, grown with various Amazon species,' Michinori said.

There were trees carrying big *cupuaçu* fruit alongside pepper plants. *Cupuaçu* is related to cacao, and is a big, ovular fruit with seeds in a white, fleshy pulp that makes a rich juice. The seeds of this plant too can be used to make chocolate. There were banana trees. He pointed out a small green bush called *gliricídia*, a legume that works as an insecticide and a fertiliser, he said, and that can help recover soil health. 'It is a good friend, a friend of nature. It protected the soil and protects the plants,' he said. In one field, African mahogany trees that had been growing for 15 years towered among cacao trees. Nearby was an area of thick, untouched forest. The farm was 'semi-organic', he said. He uses organic fertiliser and a herbicide.

He drove us out into the countryside for a long while down empty dirt roads and stopped outside the barn facing a low farmhouse with a veranda. A woman in her late forties called Francisca Ezilda Lobo served coffee and homegrown pineapple juice, and told us her husband Francisco was away on a hunting trip with friends. The previous night, someone had tried to break in and she was now sleeping with a shotgun beside her, she said. Michinori and Francisca – known as Dona (Mrs or Madame) Ezilda – had clearly known each other for quite some time. This was her and her husband's small family farm and the way she told it, Michinori had helped them transform it. 'We can't complain about our production,' she told me. 'We can live off it and live well.'

They had 52 hectares, she said, some 23 of which were used for farming and 10 of which were left as forest. They used to have around 20 head of cattle and some fruit production, but Michinori persuaded them to invest in agroforestry. So they sold the cattle and planted on degraded pasture. Now they grew *cupuaçu*, cacao, black pepper and açaí. The *cupuaçu* grew best just with the açaí, she said; the rest was all mixed together. There were native Amazon *maçaranduba* hardwood, *pequiá* and Brazil nut trees. 'Mr Michinori was always here, guiding us,' Dona Ezilda told me. 'He really encouraged us to invest in fruit growing.' He got them into the CAMTA cooperative, which, as well as selling produce, has a small factory making pulp for fruit juices and other products. That meant they always had a way to sell their produce.

We walked through her orchard and on to an area where bushes of the alien, weirdly shaped *pitaya* fruit were growing alone, bar the odd pineapple plant. It did not grow well with other fruits, she said. Francisca and her husband both came from poor families who had migrated from Ceará in the northeast, like so many in the Amazon. Her first home was a wooden shack; now she had this neat, low house with its tiled roof. For years, she had worked as a schoolteacher and one of her two grown-up children, her son Eden, had studied to be an agronomist at a federal university.

'We're very grateful to Mr Michinori, for his experience. He taught us to work, to make money,' she told me. 'We have to know how to work with our heads and our arms.' Michinori bowed his head again, and I remembered something he had said earlier. 'The best way to convince a farmer is with a farmer.' Thunder cracked and they rushed to wrap the cacao seeds drying on a 20-metre wide tarpaulin before the rain Michinori had smelt on the wind came.

Tomé-açu was not a pretty town. It was noisy and the streets were coated with red dust. There were workmen everywhere because a railway was being built, filling up barbecue joints for an early dinner. Little food carts sold typical Pará food like *maniçoba* and *tacacá*, but I wanted a table and a beer so I ate a burger in a little square. The next morning, a colourful and noisy funeral procession of cars and people went past, tail-ended by a police car. I rushed down the main road to meet Francisco Sakaguchi at the cooperative's headquarters, passing the imposing red Japanese arch of the Cultural and Agricultural Promotion Association of

Putting the Eco Back in Economy: Agroforestry Models

Tomé-Açu. Sakaguchi was the same age as Michinori, though less formal. He wore a surfing T-shirt, a CAMTA baseball cap and a wry smile. He spoke slowly and left significant pauses. He sold cacao to one of the best organic chocolate makers in the Amazon.

He told me he was a practising Buddhist but from 'force of tradition', that he did not believe that global warming had been proven to exist, and that he left 150 of his 360 hectares as native forest, from which he had only extracted some wood at different times. 'I think it's better than clearing and deforesting to transform it into what human beings call a productive area,' he said. 'I don't know if it would be productive or if one day nature would charge me for it.'

His father, Noboru Sakaguchi, had trained as a forest engineer and come on a ship to Brazil in 1957 with two agronomists, all recently graduated, encouraged by his university in Tokyo to travel and study planting rubber trees in Brazil. Working in Tomé-Açu, Noburu stayed with a family, fell in love with the eldest daughter, got married and stayed. Noburu Sakaguchi was a curious man, who brought different seedlings, such as the Taiwan papaya and durian fruit from Malaysia. And he travelled almost weekly to the state capital Belém by boat, disembarking in small riverside towns when required to, killing time by making friends with people from riverine communities. 'He was an extremely open person,' said Francisco.

On these stops, Noboru noticed that riverine people always kept a small garden where fruits and plants grew together, just as they did in the forest. The Japanese community struggled because the fungus was killing the black pepper plants they mainly farmed. He had the idea to borrow the technique and get everyone to start growing different plants together. 'His idea was to systemise this and that's where the agroforestry system came from,' said Francisco. 'It was inspired by observing the riverine people's traditional life.'

Francisco and I talked about how this tradition in turn came from Indigenous people, who have long practised their own agroforestry systems – researchers believe that areas of the Amazon forest were heavily influenced by Indigenous planting and management. His father believed that 30–35 per cent of the soil was lost through rain and agricultural exploration, and that protecting and renewing it was important. 'He said, "The native forest is very biodiverse and this biodiversity in the virgin forest is not for nothing. One species protects another,"' Francisco recalled, with his quiet smile. 'He said, "Observe nature. Observe nature and learn with her."' By the 1980s the Japanese community in Tomé-Açu was making agroforestry work.

The CAMTA headquarters was an imposing building, and the room where Sakaguchi and I talked had wooden cabinets with tiny drawers for each of its 172 members, 99 per cent of whom used agroforestry systems, its president, Alberto Ke-iti Oppata, told me. There was plenty of demand for their black and white pepper, he said. 'What we produce is not sufficient to supply the market.' They also struggled to meet the demand for their cacao and açaí – which had, he said, an 'immense market'. Most of the cooperative's members had around 50 hectares and he said preserving 80 per cent of that as a legal reserve and living off their land was unviable.

I took a bus back to Belém, pondering Michinori Konagano's conviction that agroforestry systems could work for Amazon

Putting the Eco Back in Economy: Agroforestry Models

smallholders if they had government support, training, quality seeds and the sales structure that a cooperative like CAMTA provided. One study has shown how CAMTA members produce income on areas from 10 to 20 hectares equivalent to what cattle ranchers were able to make from 400 to 1,200 hectares. 'The solution for the Amazon is the solution for the planet, which is the agroforestry system,' Michinori said. 'That is the solution I have.'

We had come to see the river giants, and we raced into the colossal, cinematically dazzling Mamirauá Sustainable Reserve on a small aluminium boat with a big outdoor motor, skating across its brown waters with barely another person in sight. When the Amazon is left virtually intact, as in places like this, its raw, magnificent nature is daunting in its beauty and scale. The technicolour skies and endless treelines dominate; humans are reduced to little dots in boats or houses. This fast, wide river and its dense forests teemed with life, and the few people who lived here did so without destroying it. This, it seemed, as the boat roared up the brown river, was the Amazon as it should be – wild and serene in a thousand shades of green.

The bioeconomy is much more than agroforestry. In the Amazon, it doesn't just work on land and in forests, it also makes use of its many rivers. I was with three men from the Mamirauá Institute – a social organisation that works in this Tefé region – to find out more about one of the most successful bioeconomy programmes here. They were making a routine visit to tiny riverside communities involved in its long-time sustainable fishing project and had agreed to let me come along. Iranir das Chegas, from a local fishing family, was with me in the back seat. Reinaldo de Conceição, piloting the boat, was from the old colonial port of Óbidos, a thousand kilometres east down the Amazon River, and Vinícius Zanatto, the São Paulo native riding shotgun up front, was looking at sustainable development.

Reinaldo cut off the Japurá River onto a smaller tributary because he had been told fishing was happening this afternoon. The slightly delayed season had begun. Reinaldo opened up the throttle and the boat roared through the breeze.

Travelling on these rivers could be dangerous. In 2017, British canoeist Emma Kelty was brutally murdered on an island further up the same part of the River Amazon we had just left. Drug traffickers regularly flew down the Japurá River from Colombia in motorboats carrying loads of cocaine and marijuana, and river pirates cruised these waters in groups of eight to ten, wearing black balaclavas and carrying automatic rifles, hoping to rob the smugglers and anyone else they could find. Holding up Colombian drug traffickers sounded like an incredibly dangerous way to make a living, so I was quietly hoping we didn't meet any pirates. Several people here told me chilling tales of seeing or meeting them, and our boat carried less fuel than was needed for the whole trip to make it less of a target.

Speeding up the smaller river, closer to the treeline, we could see life all around us. Iranir shouted out the names of the birds to me as we passed: red-eyed, grey *alencorne* birds with hooked beaks curving downwards; small, squawky *socoí*; skinny, rusty-red and black *jaçanã* with yellow beaks; devilish, sharp-beaked, grey and black *mergulhão* dived for fish; big brown and crimson *cigana* birds preened with their blue-grey faces and punk-rock plumage; red-headed *camiranga* vultures and huge white-headed *gavião panema* hawks soared overhead. Three grey-brown turtles dropped into water they exactly matched in colour, one by one, plop, plop, plop, as we neared. Black caimans the size of crocodiles lurked at muddy riverside edges, splashed into the river or glided menacingly past, eyes and noses above the water.

Our boat reached Cleto Lake mid-afternoon, just in time to catch the hunt for the *pirarucu* fish. These river giants can reach two metres long and weigh up to 300 kilos. Two dozen fishermen in canoes had formed two circles with nets, surrounding the fish, and they were picking them off, one by one, shouting and bantering as they did so. As the circle of net closed, one man stood at the front of the canoe with a long harpoon and speared the enormous *pirarucu* – each as long as a man was tall. Others dragged the fish out of the water and bopped them to death with sharp blows from a wooden mallet in exactly the right spot on the forehead – the sound of the blow was like someone whacking hollow balsa wood. 'The guys are good at their art. They can hit the fish freehand,' Iranir told me. The remaining fish splashed violently as the net closed in.

We were outside one circle of nets and it was suggested I get a closer look. I stepped hesitantly into a narrow canoe with fisherman João Cordeiro Neto, who sat calmly crosslegged under a straw hat while I perched nervously behind with the nets and ropes as he paddled closer, worried that one of these pink and grey fish might capsize us in panic as their death approached – this apparently does happen. The men dragged the fish they had killed into a bigger boat with a tarpaulin roof. They took 64 out of this lake this day, and 578 over 13 days.

Laid crosswise, the *pirarucu* were longer than the boat was wide, and their flat snouts and enormous gaping mouths with tiny, saw-like rows of teeth hung over the edge. *Pirarucu* have a long, narrow body and a flat rear. Only the young breathe mainly with gills; adults breathe air through a swim bladder. The *pirarucu* is the largest scaled freshwater fish in the world.

Its rich, tasty meat is widely prized in Amazon countries and has traditionally been an important source of protein for many people in the region, but overfishing reduced their stocks and catching them is now only allowed in areas with special authorisation and a sustainable management plan, called a *manejo*. Here in Amazonas State, *pirarucu* fishing is allowed in 34 areas, home to 305 communities and 11,000 fishers.

Mamirauá Sustainable Reserve has long had a successful *pirarucu* programme. Launched in 1999, it is run with the Mamirauá Institute, a non-profit research and conservation organisation set up the same year in the bustling little town of Tefé, on the banks of the river of the same name. The Mamirauá Institute is supervised by Brazil's Ministry of Science, Technology and Innovations, and has a smart campus and research centre that does not let anyone enter on a motorbike without a helmet. Their centre has a rich collection of stuffed Amazon wildlife and archaeology, a rare sight in a remote Amazon town. It works in the state government-controlled Mamirauá Reserve, which covers 1.1 million hectares of forests and rivers of *várzea*, or flood plain, and the Amanã Sustainable Reserve, which is twice the size and right beside it.

Less than half an hour back downstream from Cleto Lake was a floating platform with wooden rooms and boats tied up alongside. Wearing hair nets, rubber boots, white uniforms and

Putting the Eco Back in Economy: Agroforestry Models

masks, the fishers dragged the enormous fish out of the boats and loaded them on a conveyor belt. In teams of two, they sliced them open, cut out their gills and organs and washed them out with a hose. Blood gushed across the floor. Two women counted the fish as they were cleaned and put on ice on a boat. It went remarkably quickly, this processing line of giant fish in the midst of a wild forest, hours by boat from the nearest town.

Milce Carvalho was the only fisherwoman in the group. Once the fish had been cleaned, she stood on the open deck at the front of the platform and watched two men load all the gills and guts and paddle them over to the other side of the river to dump them on the riverbank. I counted seven big caimans lurking. 'They come around this time. They already know!' Milce grinned. A flock of vultures descended and the caimans began gliding towards the pile of bloody, sloppy guts. One huge caiman lumbered up the bank towards the offal and the vultures fled. Milce and I laughed at the jungle drama playing out on the other side of the river. Her four grown-up children lived in a house in Tefé she had bought. 'I didn't have any savings before the reserve began and the Mamirauá Institute started working with us,' she said. 'I'm happy I'm taking money from here for what I need to have, a good house for my children.'

Wearing a pink T-shirt and a straw trilby, Raimundo Rodrigues told me the battle was getting a decent price for the precious fish – which hovered around four to five reais per kilo Locals knew they deserved more from the traders. In Rio de Janeiro, thousands of kilometres to the south, a kilo of *pirarucu* sold for nearly 70 reais. 'If it wasn't for the institute's encouragement, we wouldn't have this reserve here today,' he said. 'This is our home here, it's us who take care of it, us who decide what we are going to do.'

The platform, which the Brazilian government financed for this sector of the reserve, had solar panels for power and recycled its water. In a small kitchen, a group of women cooked chicken, *pirarucu* bladder, rice and beans. They laid out the food in pots on a blanket. It was swelteringly hot. Night fell and Iranir played *carimbó* on the guitar – a swaying musical style from the eastern Amazon. We slept in hammocks strung around the open-air area where, hours before, the enormous fish had been unloaded. I woke up several times in the night, thinking about those enormous caimans just a few metres away. People told me they didn't attack, but they also told a story about a woman from the Mamirauá Institute dragged underwater by a caiman while washing clothes years ago. The woman survived but lost a leg.

After a noisy breakfast of chatter, crackers and coffee, Reinaldo started the motor and the aluminium boat roared off up the river. We passed a riverside hut where three men and a woman were roasting manioc flour – a carbohydrate staple for Amazon people. There was a splash in the river as a grey *tucuxi* dolphin dived, then, minutes later, a flash of pink as a freshwater *boto-vermelho* dolphin did the same. We sailed on. Above us on the bank was another community. 'This is *terra firme*,' said Iranir – that is, forest on dry land, rather than the flood plain, or *várzea*, of Mamirauá, and was actually part of the neighbouring Amanã reserve. The community of a few dozen houses was called Curupira. Iranir said it was named after a spirit, the animal of the forest.

We moored up and climbed a steep, grassy incline – a good 45-degree angle – to the wide, open porch of a wooden house.

Putting the Eco Back in Economy: Agroforestry Models

A group of people were poring over the Curupira community's *pirarucu* sustainable management plan. For the next couple of days, Reinaldo, with his laptop, and Iranir joined them. They examined charts and tables and argued over the accounts of the operation we had seen the previous day, discussing new options of who they could sell the next catch to.

The family house was made of blue-painted wooden slats with a corrugated iron roof. There was a wooden toilet shack over a pit out back. There was a cooker and a kitchen but also an open-sided room down some steep wooden steps with a tank of brown river water to scoop over your head for washing, and a hearth for cooking. One afternoon, I went in to bathe and found half a dozen fish laid there, including a *piranha* with its angry, horror movie teeth, strung along a pole shoved through their mouths. Just move them aside, said host Francisco Costa. The fish that day was especially delicious.

The group moved around on the terrace as they worked through the details: democracy in action is slow and laborious. There weren't that many places to sit – a bench by the table, a few low planks and Costa's favourite, a short piece of wood placed over the metal ring of a large gas bottle. Washing was strung out to dry around the terrace, which had a glorious view over the river and forest. There were some pigs in a wooden stockade, a puppy and an angry, screeching pet parrot. People came and went. When lunch and snacks were served, everyone sat on the floor to eat.

Fishers in Mamirauá told me they mostly lived from other fish they caught to eat and sell, like *tambaqui*, a big, rounder fish, smaller than *pirarucu*, that is also highly prized. Fishers get a minimum wage from the government not to fish during the breeding season, and many here received a government cash transfer scheme for low-income families, called *Bolsa Família*, or Family Allowance. Families also got a R$100 (US$18) a month Guardiões da Floresta (Forest Guardians) payment from the Amazonas state government.

The sustainable *pirarucu* programme gave those taking part a vital chunk of extra cash that people said enabled them to save to buy things – like a new boat engine or the house Milce saved up for – that would normally be out of their spending reach. But with it came a heavy commitment. The 83 fishers in this sector had to constantly monitor the lakes where the *pirarucu* mostly live for invading fishermen, putting aside money from their catch to pay for fuel. They had to count the fish before harvesting them, forming lines and looking at what was there and what size. Each fisher had a quota and those that slacked in commitments – for instance, not taking monitoring shifts or paying monthly subscription – had their quota reduced. When they found someone fishing illegally, they confiscated nets and equipment and expelled them.

The fishers ran the scheme, Reinaldo explained. He and Iranir were there to give advice and some guidance. 'The people who do this work should be more valued,' he said. This, he argued, was how to protect the Amazon. 'I believe that sustainable development is possible,' he said. 'This is the way.'

Many of the people here were descended from rubber tappers who had moved from the northeast during the Second World War rubber boom – when a US government programme helped pay for tens of thousands to move – and afterwards. Maria das Graças – one of the fishers in the group, with her husband Vernior

Putting the Eco Back in Economy: Agroforestry Models

Batalha – told me her father had worked for a boss who had the monopoly on buying his rubber production and selling him overpriced goods. This system, now seen as a kind of slavery, was widely used in the Amazon and many rubber tappers were kept in perpetual debt. Tappers were murdered for failing to meet their rubber quotas. 'You had to find a way to pay by the end of the month,' Maria said. 'The boss was a bit of a tyrant.'

The *pirarucu* project had been running for more than a decade. Before then, it was 'pretty bad', Francisco Costa told me one night, sitting on his son's bed while the TV boomed in the next room. Outsiders overfished *pirarucu* and other fish until stocks were dangerously low, and fishers like him were 'persecuted' by environment agencies because they fished illegally. 'It's hard for men to control things. Men abuse the abundance that God gave,' Francisco said. 'Men need controls, otherwise they finish things off.' Maria said she became environmentally aware when she took part in a Catholic study group and later became an unpaid environmental agent for Ibama. 'Today, there is real abundance here because of our work,' she said.

Francisco and his wife, Lucineia de Sousa, had offered us this bedroom used by their two teenage sons, but we slung hammocks on a beat-up, pale blue, two-storey boat. A large and ominous spotted frog slept in one cranny I steered well clear of, after being told it might be poisonous. At night, people swapped stories. A young man called Jonas Oliveira recalled the rare black jaguar that had terrified Curupira village a year earlier. It killed dozens of chickens, as well as dogs and cats, before villagers shot it dead, and it was still twitching with life after taking three shots. 'A jaguar is a dangerous animal,' Jonas grinned.

Curupira was very still, beautiful and timeless, except when somebody started playing pop music or Evangelical Christian songs at high volume – but that's the same across the Amazon. It felt like the future and the past existed in the same place, that people and nature were part of the same thing, not in an eternal battle with each other. Time and people and boats on the river moved slowly. The overwhelming sense of vastness was both calming and a little unsettling. And people saw their world differently from me, as I found out one afternoon when Francisco Rodrigues, one of the fishermen in the project, took me upriver to show me his plantation – the '*roça*' that riverine people use to grow fruit and staples like manioc.

After just ten minutes in his little canoe with its chugging motor on a long pole, we were in another world. We sailed between curtains of trees immersed in water while a bird fled, calling plaintively, and emerged into a wide, peaceful, lovely lake surrounded by flooded forest called the Largo do Pintado – the 'Lake of the Spotted One' (or 'Spotted Beast'). An invisible old man was supposed to live there, Francisco explained, and an enormous snake. 'Lots of stories here in the Amazon,' he said.

As we got out of the canoe, he told me more about the forest spirit or creature that had given the village its name. Belief in the *curupira* is widespread across the Amazon, though I had always struggled to grasp what it actually was. A folkloric Amazon figure whose story was shared to the outside world by a Jesuit as early as 1560, the *curupira* is described as a forest spirit that protects the forest, its animals and trees, ensuring those who seek to damage its nature get lost. The name comes from the Indigenous Tupi-Guarani family of languages and means 'boy's body'. Though

Putting the Eco Back in Economy: Agroforestry Models

Francisco called the *curupira* 'she', I noticed. But then again, everything in Portuguese is either male or female – a snake or a table is also 'she'.

Standing under the trees, far from any other humans, with the wind rustling the trees, Francisco's story took on a spooky resonance. 'It's a forest animal, *curupira* is from nature,' Francisco said. The *curupira* was the height of a man's shoulder and her feet were turned backwards, so she used her forward-facing heels to beat out a rhythm on the bases of the trees while whistling softly and mournfully: 'Whooooo! Whooooo!' Francisco demonstrated the sound of those heels beating on the tree – 'toc, toc, toc' – and his *curupira* whistle mingled with the breeze. He said he had seen the *curupira* many times.

'We've seen her moving over there by the stream. We have seen her looking at us. But it's very fast animal. They spy from behind these trees and hide and nobody sees them again because they are invisible,' he said. 'She disappears.' The *curupira* caused no harm, he said, but there was a curious detail – and this explained why he called it 'she': the *curupira* only appeared in female form. Nobody ever saw the male. Old folks said its male was the *tamanduá bandeira* – a giant anteater. 'It's interesting huh? Interesting beast,' said Francisco.

As I processed this mind-blowing information, Francisco led the way up a short slope to his plantation – an allotment, cut and gouged out of the thick forest – where he was growing avocado, lime, banana and manga. To an outsider, its mess of trees, bushes and plants might not have looked that different from the forest itself, but this was a valuable source of food. He pulled aside a tangle of vegetation to show me the manioc growing underground like a potato. This wonder-root and staple has a myriad of uses as a delicious carbohydrate. It becomes dried flour, *farinha*, used to bulk up meals, and tapioca to make savoury pancakes; it can be cooked like a potato, roasted or fried; Indigenous people make it into a hard, flat bread called *beiju*, and it also makes *tucupi*, a tart yellow sauce to cook fish in. Some kinds contain poisons that have to be removed before they can be eaten. Francisco called the manioc '*roça*' too, as if his allotment and the vegetable that provided a carb-load were one and the same.

Not everyone in Curupira village took part in the *pirarucu* programme – half a dozen or so people did not. I went to see one, and the story he told me seemed to say a lot about how dangerous some of the Amazon's illegal, yet widely tolerated activities could be.

Daniel de Sousa was 24 and lived with his family in a wooden house with a steep wooden staircase. He was a fit, handsome guy with a black beard and a young family. He had stopped fishing after a 'problem happened', which he didn't explain. Instead, he farmed. The previous year, Daniel had been a *garimpeiro*. He had joined the team working on an illegal gold mining barge upriver. He worked with four other men and a woman cooked for them. His job involved using a chainsaw to clear forest so the barge could enter into a river on a protected reserve. He had planned to use the money he made to open a store on the riverside. But after four months, he left.

'I don't really like trouble and it's a serious business there, there's death; death happens, everything happens in the *garimpo* and only us that see it know,' he said. 'There's a lot of prostitution, and drink, drugs, you have all kinds of drugs there.'

Putting the Eco Back in Economy: Agroforestry Models

A bigger boat, moored up near his barge, housed a *cabaré* – literally this means cabaret or nightclub, but in this context it means a brothel and bar. *Garimpeiros* get paid in gold and there is a rudimentary yet efficient service industry around them. Shopkeepers, bar owners and pimps – often one person combines this role – make good money out of *garimpeiros*, charging vastly inflated prices in gold to supply what they need in remote jungle locations. Daniel, like his wife, swinging with a baby in the hammock behind us as we sat on stools, was an Evangelical Christian and did not drink or smoke. (I didn't see anybody in the reserve drink or smoke.) But the cabaret was also the only place with an internet connection and so he went there a couple of times to call home.

'That's where I saw this business, one guy killing the other,' he said. He counted 14 stab wounds in the vicious attack. There was no medical care. 'It was so cruel, there's no pity,' he said. Daniel also saw cellphone video of a horrific accident on a neighbouring barge when a loose, flying cable killed two brothers – one was hit in the neck; another lost half his head. He also heard about an older woman murdered while she slept on a boat that was also a cabaret, apparently to steal her money. Violence, drugs and prostitution are common in isolated *garimpeiro* areas. But the five or six deaths that happened while Daniel was there were enough to scare him and send him back home. He thought the work should be legalised and controlled by police. 'There is a lot of gold there,' he said.

Daniel only had a vague notion of the damage caused both to *garimpeiros* and Amazon communities by the mercury used in illegal mining. Nor was he very clear on what climate change was: to him, it was something talked about on Brazilian television news programmes he rarely watched. I tried to explain it to him. 'Um-hum,' he said, uninterested. Protecting the forest was good, but people also had to have jobs – and jobs around here were hard to come by, he argued. He had planned to go back upriver and get the half of the salary he had not been paid, but changed his mind after hearing about more violence – this time apparently involving Colombian guerrillas coming over the nearby border. Months later, a Brazilian TV report interviewed the widow of a *garimpeiro* in the same region whose barge was invaded by Colombian guerrillas – she was forced to cook for them and her husband taken away and later killed in a shootout. According to Brazilian authorities, Colombian guerrillas have been trying to dominate the illegal extraction of gold. In 2022, a study showed that *garimpo* had destroyed 531 hectares of the reserve Daniel had been working in since 2019.

Daniel was owed 145 grams of gold, or around 30,000 reais, he said. He seemed to have given up on ever getting it. 'It wasn't God's will,' he said, as his small children played noisily. 'We have our plans and God has his.'

Our noisy aluminium boat stopped at other communities on the way back to Tefé. We had coffee on the veranda of the house where Raimunda Meirelles lived, overlooking a spectacular river panorama. She was in the kitchen washing fish, a sharp, quietly spoken and sympathetic woman in her late fifties who had been here since 1982 and loved it. 'There are fish here, everything is free,' she said. Raimunda had five children and ten grandchildren and was the coordinator for the fishing project in this sector. She told me how monitoring shifts on their *pirarucu* lake worked. The

Putting the Eco Back in Economy: Agroforestry Models

fishers spent three days on the floating base there, teams of two, three or four people. It was arduous and dangerous persuading invading fishermen to leave. Some went quietly, others became angry and threatened to kill them. 'It's not easy, no, at times we feel the knife on the neck,' she said.

Like Maria das Graças, Raimunda was also an environmental agent and also had been given a training course by environment agency Ibama. Agents like her have no power; they cannot levy fines and they work for free – their work is mostly educational. The unpaid work got more difficult after Ibama closed its office in Tefé and stopped patrolling the reserve. 'If we could advise everyone to do what has to be done it would be much better. But there are still people who don't understand this and we always find it difficult to make people understand,' she said. 'Ibama left and us environmental agents were left without a father.'

Raimunda had a very good reason why the forest should be protected – because it cools everything down, something that becomes instantly clear to anyone who has ever walked out of glaring Amazon sun into the trees. 'If there was no forest, everyone would be finished because the climate is very hot,' she said. Since we spoke, new research has confirmed what Amazon residents like Raimunda Meirelles have always known: that forests keep the Earth at least half a degree cooler, helping to keep air moist. In the tropics, the cooling effect of forests like the Amazon is more than one degree. Like so many other Amazon residents I spoke to, Raimunda was feeling climate change in the heat of the sun on her neck. 'Before we used to go to the *roça* and work all day,' she complained. 'Now we can't stand it. At 9am, when the sun is out, we can't stand it any more.'

I wanted to know how and why people living here in the reserve thought it was so well protected, given how grandiose and natural it seemed, and how that could be maintained and improved. The answers seemed revealing in both their simplicity and good sense.

We visited João Cordeiro Júnior, son of the man whose canoe took me closer to the *pirarucu,* in his community of floating houses called Nova Jerusalém. His was a vision of collective responsibility and individual leadership. A taut, energetic, smiley man of 26, he told me that becoming a fisherman was a family obligation – he had 13 brothers and sisters. 'It's like there's no choice for us,' he said. He valued fishing with his father, brothers and other relatives. 'To protect the Amazon, you have to have a leader in the community,' he said, describing their network of association presidents and vice-presidents. 'There is a whole network there that discusses the best way to preserve the Amazon.'

The final stop was a village up a muddy creek called Jurupari – not a part of the reserve but connected to its sustainable fishing networks. There, in a little wooden house on stilts belonging to Maria Cecilia da Silva – a voluntary midwife, among many other things – and her husband Raimundo Gomes, both in their sixties, we sat on the tiny kitchen floor and ate a stew made of a fish called *bodo*. It had a tough, spiny, spiked armour you had to take great care not to eat as it could stick in your throat and choke you – which did not deter the toddler sat beside me wolfing it down.

All the houses were on stilts because of seasonal flooding, and a gang of children ran around the village's grassy fringes roaring with laughter in what for them appeared to be a giant, car-free kids' playground. I asked Romário da Silva, a sharp, articulate teacher in his twenties and a fisherman's son, what he thought was the best way to keep protecting this place. He advocated more help. 'Support from the environment ministry and local government. Today we feel we're alone in preservation here, the only partner we have here is the Mamirauá Institute, which helps us a lot,' he said. 'Partnership, right? Fundamental.'

Belém and Manaus

CHAPTER FOUR

Stop Destructive Development: Managed Urbanisation

Dom Phillips
Manaus, Amazonas

'Art has the power to turn the key in your head.'

David dos Santos, just 17, was lying spreadeagled on his back in the dirt when we got there, a crowd standing around his dead body. Bystanders said he had tried to run when armed men appeared in this alleyway in Mutirão, a poor neighbourhood in the Amazon metropolis of Manaus, one Saturday evening. An aunt said he frequented an Evangelical church; a neighbour said he had previously been involved in the drug trade. In a place like Mutirão, where vicious rival gangs vied for control, nobody I spoke to wanted to give their name.

David's relatives grabbed onto each other for support, wailing and keening as his body was lowered into a metal tray and hoisted into the morgue truck by a man in shorts and a police forensics officer wearing a black cowboy hat and a large knife in his belt. Forensics specialist Isabella Erthal later said he had been hit in the back by one bullet that broke his clavicular and exited his chest. One local media story called it a stray bullet. Another said the killers were looking for someone else, but the teenager was in the same group and ended up being the one shot dead.

Nayara Felizardo, the Brazilian journalist I was reporting with, and I had wandered onto the street when suddenly there was a flurry of movement and burst of screams as people fled the sound of a gunshot. We ran too and crammed into another reporter's car. A woman, one of David's relatives, came to the passenger door window. 'Call the police. They are invading,' she said.

The two reporters in the front fumbled for a phone, trying to call the police. Nayara was scribbling furiously in her notebook. Josemar Antunes, the local crime reporter we had come to meet, got through to the cops. 'There has been another shooting here,' he told them. Behind us, a hundred metres or so away, a black car appeared, its headlights on. It stopped. Time went very still. I thought about how vulnerable we were right then, how it could just pull up alongside us and a gunman inside it easily riddle our car with bullets. And I thought about how I was the one who had brought us both here. The car passed us slowly. 'I always feel nervous when I see a black car like that,' said Josemar. 'See how long it takes? If this was the United States, the police would be here by now.'

We stayed in the car for about ten tense minutes until a police light appeared just down the road and Josemar said it felt safe enough to get to his car and get out of there. That was an important lesson on the crime beat, he told me: get away from the crime scene as soon as the police have gone, if not before.

Manaus has so much drug crime – and so much interest in it – that it has a thriving media network to cover the violence. This was immortalised in a Netflix series called *Killer Ratings*, about a local politician called Wallace Souza who hosted a crime TV series that always got to the murder scene first – and was then accused of masterminding the very crimes he reported on.

Stop Destructive Development: Managed Urbanisation

Manaus is 1,200km from the sea, where two major rivers meet: the Solimões (the upper reaches of the River Amazon) and the River Negro ('Black River'). It is the biggest city in the Amazon with a population of around 2.1 million people. The first European settlement was a fort built in 1669 and the city has gone through a series of ups and downs since then. Between 1890 and 1910, Manaus boomed along with the rubber trade. Its ornate opera house, the Teatro Amazonas, was built, and Manaus was one of the first Brazilian cities to have electricity and a telephone system. The rubber barons became so rich that some, it was rumoured, sent their laundry to Europe.

But the rubber industry depended on wild rubber trees and in 1876 a Brit called Henry Wickham smuggled 70,000 seeds back to England. Within a few decades, rubber plantations in Sri Lanka and Malaysia had finished off the Brazilian rubber boom. Manaus only really began to recover after it was made a free trade zone in 1967. Manufacturers flooded into the city and now produce consumer goods like cellphones, televisions and motorbikes. The tax breaks make manufacturing 35 per cent cheaper here than in the rest of the country, according to Ernst & Young, the professional services firm.

But that has not made the people of Manaus any better off – quite the opposite. According to Brazilian government statistics agency IBGE, 38 per cent of the households in Manaus earned the minimum monthly salary of R$998 (US$247) or less, ranking it at 3,021 out of the 5,570 cities in Brazil and lowest of all the 62 municipalities in the sprawling state of Amazonas. According to a study by the Amazon 2030 network, workers in the Manaus industrial zone got 5 per cent of companies' turnover, compared to the 11 per cent workers are paid on average nationally.

HOW TO SAVE THE AMAZON

For towns and cities across the Brazilian Amazon the levels of general infrastructure services like garbage collection, power and sewage are worse than towns and cities of the same sizes in the rest of Brazil. This became pretty evident to me on this trip in 2018, when I spent three nights driving around Manaus's poorest neighbourhoods with Josemar. It was like a crime tour. And Josemar, a no-nonsense, hard-talking reporter brimming with grisly tales, made for an excellent guide. Every corner seemed to spark a story of a shootout or gruesome killing. That was where the body was found in a suitcase, he said, gesturing at a grassy verge as he put his foot down to keep up with the racing morgue truck, constantly checking his cellphone messages. Josemar was fast and efficient and he knew his beat well. He seemed to thrive on adrenaline, roaring around in his beaten-up old car as he recounted gang sagas. Warring drug gangs were behind most of the killings, he said.

On New Year's Day 2017, 56 people were slaughtered in a brutal prison massacre that sent shocked headlines around the world, after the local Northern Family drug gang (Família do Norte, FDN) attacked prisoners from the rival First Capital Command gang (Primeiro Comando do Capital, PCC), South America's biggest drug gang, from São Paulo. Five days later, at least 30 prisoners were murdered in what was reported to be either a revenge attack or an internal conflict within the PCC at a prison in Boa Vista in nearby Roraima state. The violence mushroomed. I reached out to police officers and security experts in Amazon towns, and my WhatsApp quickly filled up with gruesome videos of decapitated heads, mutilated bodies and even a heart left on a prison corridor floor. Other Brazilians began sharing the images; so did the gangs. A cop in Manaus sent me a video of a wild-eyed man just before his head was sliced off by a machete. A video of the Roraima prison killing spree featured a sharpened metal instrument being shoved into the eye of a screaming, wounded man.

In October 2021, I went back to Manaus and met up with Josemar. He was still working the crime beat, but kept leaving or getting fired from one news portal and joining another. Now in his mid-forties, with five children including a four-year-old girl with his latest partner, Eliane, a supermarket cashier, and after 12 years chasing crime scenes, he was feeling the strain, he said. Josemar used to work for an electronics factory before becoming a crime reporter, and he had always been fascinated by police work and crime. He used to watch Wallace Souza's TV show avidly.

'I wouldn't say I was hooked on it but I always liked the idea of reporting the accidents and crimes of murder,' he told me, sitting outside the police homicide department as traffic growled past. While some reporters baulked at grislier crimes, he claimed to be unaffected. 'I've covered many deaths here in Manaus,' he said. 'It never messed me up.' Even confronted with putrefying corpses, while colleagues vomited, he declined to use a mask. 'To me, that was just another environment,' Josemar said. His colleagues had given him a nickname: 'The Prince of Death'. He liked it.

Josemar took a hard line on crime. A friend who worked in the morgue had lost his son, shot dead for painting 'FDN' on a lamppost by gang members from the Red Command (Comando Vermelho, CV), a Rio de Janeiro-based gang, he said. The father swore vengeance and was also killed. Josemar lost a nephew, who got involved in the drug trade and was later killed by rival gang members. Another nephew was monitored by an ankle bracelet and a third had drug problems. 'It's difficult,' he said. 'It was sad.'

Josemar said the drug trade was what drove the violence. 'The whole of Brazil is a violent country today because of the drug trade,' he said. 'What makes me sad, in a crime situation, is when I saw that a child was abused, that messes with me, or an old person who was run over or murdered by a stray bullet. That messes with me because I see them as vulnerable people.'

People involved in the drug trade, that was something else. Josemar had never even smoked a cigarette, never mind experimented with any drugs, he said. He was focused on the story. And Manaus was getting even more violent – in 2019, 657 people were murdered here, but 2021, that total was 1,060. In ten hours in May 2022, ten people were killed. Now the gang dynamic had changed, Josemar said. The FDN was seriously weakened and the Comando Vermelho were the dominant gang.

Some reports said the Comando Vermelho had been involved in another homicide we had reported on in 2018. We had ended up deep in another poor area called União, walking alongside wooden shacks thrown up beside a polluted canal, where a man called Carlos Bruno Miranda had been shot at least seven times. Wearing a pink T-shirt with jeans and Adidas trainers, the victim lay slumped outside a shack, covered in blood, a piece of paper on his body with the names 'Fabio' and 'Regis'. Miranda was 26 and had been wearing an ankle bracelet. He had been released from prison while serving a five-year sentence after he and four others had robbed a couple at gunpoint and stolen their car.

A crowd had gathered around Miranda's body. Two of his friends had come to carry it away. 'I think he used cocaine but he stopped,' one said. 'He was involved at 18. He had an ankle bracelet for stealing a car. He spent a lot of time here.' They thought

Stop Destructive Development: Managed Urbanisation

his death resulted from a 'settling of accounts'. Miranda's friends put on plastic gloves and pulled his bloodstained T-shirt over his face. A TV reporter went along, broadcasting live. I followed with Ney Gama, a police officer I had talked to earlier at the Homicide Department HQ, now wearing black combat gear. Gama had told me the police struggled to control the rampant violent crime. 'The first thing is a lack of people, and then the lack of equipment and structure,' he said. Now, as this macabre procession made its way alongside the shacks with their glassless windows and satellite TV dishes, he showed me photos of his daughter. 'Too many homicides here,' said a woman as the body was loaded into the truck.

As a city, Manaus seemed to have something of an identity crisis. It felt like it had really never wanted to be located in the middle of a jungle, which was why its richest patrons insisted on building a European-style opera house and drinking French champagne. That impression was confirmed when I visited the Galeria do Largo – the Gallery in the Square, right opposite the opulent Teatro Amazonas. Upstairs, the permanent exhibition at this state government-run gallery was Santa Anita City, a miniature town created by Mário Ypiranga Monteiro over 50 years ago. Intricacy realised, Santa Anita City reminded me of the creations of model railway enthusiasts, with its neat streets, little figures, buses, cars, road signs, houses, churches and, of course, railways.

It certainly was impressive, if a little unsettling in its attention to detail. But two aspects of this work, which filled an entire room, stood out. One was that it looked nothing like Manaus now or then; instead, it resembled a midwestern US city. The other was that rather than showing the lush, exuberant rainforest that surrounds Manaus, and despite a miniature Brazilian flag, Monteiro had elected to recreate pine forest, snowy hills and even a ski slope. The closest reference to the rainforest I could see were mini tractors carrying away tiny logs.

Downstairs, the gallery had made up for Monteiro's incongruous oddity with an exuberant and vivid contemporary exhibition depicting Indigenous scenes and political messages, such as a graffiti image with the slogan 'demarcation now!', which calls for more Indigenous territories to be created. A young woman and a young guy were browsing the exhibition and taking selfies in front of their favourite paintings. They had no idea what was upstairs, even though they had been here before, they told me, but they really liked the colourful works about Indigenous life and culture.

'I like the way it's arranged; it's something modern, so it's much more able to attract the attention of young people,' said Amanda Magal, who was 24. 'It's able to be really emblematic in this space.' Vitor Silva, 23, liked the way it showed 'our culture, our daily life, our Indigenous roots', he said. Santa Anita City suggested to me that the development model for Manaus had been skewed from the beginning. It hadn't ended poverty and it hadn't prevented young men in poverty from joining gangs for easy money – more than what a tax-free factory was paying.

'Preservation is something we want and we aim for in our day to day, with basic things, with water, reducing this and that, but it's a much bigger issue,' said Amanda. 'The Amazon issue is in the hands of politicians; it always has been.'

On a hot August morning in the Amazon city of Belém, I followed Renato Rosas and two colleagues carrying large bags of disposable diapers for donation through tight, constricted alleyways that led

Stop Destructive Development: Managed Urbanisation

between the cramped blockhouses, all raised on stilts above the River Guamá. Everything in one of the poorest, most dangerous sectors of the sprawling Jurunas favela had been constructed from planks of wood – perhaps why these cramped alleys made sharp, right-angled turns. The little houses were so close together you could see inside and there was no glass in the windows. Cables and wires hung low around our heads. I could smell the river and the sewage below. There were shops and stalls selling açaí, a basic food for so many people because of its low price and high nutritional value – here it is eaten with fish, chicken and meat. Renato pointed out plastic buckets of açaí stones that could be used to make bricks – another of his planned sustainable projects. People sat on tiny terraces. A motorbike eased past. 'It is the most extreme poverty,' Renato had said of the favela a year or so earlier, when I first talked to him.

A musician and biomedical salesman turned community leader, Renato has a powerful personality. You can see there is a performer in him. And perhaps a politician, given his obvious popularity here in the alleyways. He was greeting people the whole time, shouting out jokes and 'hey how is it going, *beleza?*' – a slang greeting in Brazil which literally means 'beautiful' but is understood as 'all good?' He had a word for everyone. He told me later I had met the drug gang member on watch. One young guy he also chatted to was the son of a celebrated local musician, now working for the city hall, who had a team there collecting people's names to ensure they were getting benefits.

There are favelas all over Brazil, improvised jumbles of poor housing and people, where culture and community life is vibrant and noisy, the influence of local or federal government is minimal, drug gangs set the rules and sell their wares, and violent police incursions against gang members are commonplace – as are the deaths of innocents from stray bullets. In Belém, favelas are called *baixadas* – lowlands – and with around 60,000 people living here, Jurunas is one of the biggest in this city of 1.3 million. Belém has a long colonial history, which can be seen in some of its stately buildings, noisy and busy traffic, and a famous riverside market that begins throbbing with life and people and produce in the early hours of the morning.

The city – whose name means Bethlehem in Portuguese – also has plenty of dirt-poor suburbs and favelas. I had visited some years earlier, reporting on the massacre of ten young men, executed by a convoy of men in black after a rogue cop known as Pet, who ran a paramilitary gang, was killed. The state assembly produced a detailed report on paramilitary gangs involving police officers, called *milicias* in Brazil, soon afterwards. Here in Jurunas, we were less than three kilometres from the upscale neighbourhood of Nazaré – Nazareth – with its quiet, wide streets and expensive coffee shops, but we could have been in another universe. People living here only go to places like Nazaré to work as maids to richer Brazilians or employees of upscale coffee shops and restaurants.

Both in their forties, sisters Lucia and Lucilene Dias had worked as maids but lost their jobs during the pandemic. Renato and his colleagues had come to donate the diapers for their mother, Maria, who was 89 and had been bedridden for five years after a stroke. She lay on a bed in the downstairs area, which included a kitchen, lounge and shower, fully visible through the open window

Stop Destructive Development: Managed Urbanisation

to anyone outside. A brother, his partner, Lucilene's children and a granddaughter also lived there. Nobody worked, they scraped by on the Bolsa Família benefit Lucilene got and odd jobs. 'Nobody knows what we're going through, only God,' Lucilene said. 'A crisis.'

When the Covid-19 pandemic hit Brazil like a sledgehammer in March 2020 and cities locked down, people living in favelas like Jurunas were left without income and, in many cases, food, while the government and Congress grappled for weeks over an emergency payment scheme. Civil society stepped up, with favela non-profit groups and associations raising money to buy parcels of basic food, cleaning and hygiene goods, and then delivering them. Renato Rosas, then 38, had a music and education project called Farofa Black that became one of the food distribution networks. I spoke to him by cellphone at the time, while writing a story about these self-help networks springing up in Brazilian favelas. He sent photos of the food distribution in the *palafitas*, as these wooden shacks on stilts are called, and of a huge sururi snake, a kind of anaconda, curled up in somebody's bedroom. These were common, he said.

Now his organisation OCAS, or Community Organisation of Social Adhesion, helped residents with everything from job training to education to deliveries of necessities like these diapers. The donation of diapers was 'very important, really' said Lucilene. Both women attended Evangelical churches. But while Lucilene supported far-right president Jair Bolsonaro, her sister Lucia did not. Nor was this their only difference of opinion. While Lucilene said she had never heard of the global debate about conservation and development in the Amazon, Lucia knew about it from watching television. 'I don't think the Amazon should be deforested,' she said. 'I think it's wrong.'

I asked her what sort of impression she had of Europe, of England. 'I see it on the television when it's on. Hmm. It's beautiful, huh. It's different to Brazil,' she said. Very different, I said. 'Basic sanitation, mainly,' she added. And here? 'There practically isn't any. Garbage in the street, open sewers, drains,' she said. To me, her reply was a tiny glimpse into the huge challenges involved in living in a community like this. The first thing that came to mind when she thought of England was a proper sewage system and running water. Theirs had to be pumped up from the river by an electric pump they could barely afford to run.

Earlier that morning, I had watched a small group of Jurunas residents as they listened quietly to Ana Claudia Conceição in an open riverside space that normally held live shows and discos. The space was right beside the river, with wooden walls, potted plants, a mezzanine and a corrugated iron roof, next to a boat workshop. Ana Claudia was explaining how OCAS had been set up for education, to train employed and unemployed people, but had ended up delivering '*cesta básica*' – food baskets to residents during the pandemic. She wore a yellow T-shirt with the OCAS name and worked for a state government-run university, but had lived her whole life in the neighbourhood. 'I always liked to help people,' she said.

A warm and serious woman, Ana Claudia exuded capability, tact and good sense. She had worked for 15 years in a community centre but was attracted to Renato's project because it aimed to educate people. 'Its main objective is to train residents with

Stop Destructive Development: Managed Urbanisation

courses, workshops, lectures, so we can try our own businesses,' she explained. 'They're going to need to be trained to get back into the labour market.' Watching Ana Claudia talking, I was struck by how important grassroots organisations like hers are to disadvantaged communities, where the state's presence is rarely felt, education levels are often low and people have been trampled on by society since they were born. The dozen or so residents listening seemed to feel that. They paid attention and all, unusually for Belém streets even in August 2021, wore masks.

There were banners behind her. One read: 'Jurunas, our neighbourhood, our home.' Another said 'G10 favelas'. The G10, which took its name from the international meetings of rich countries, is a national network of favelas set up by Gilson Rodrígues, an activist and community leader from São Paulo's massive Paraisópolis favela. It held its first meeting in November 2019, which Renato attended. Rodrigues and his colleagues in Paraisópolis led the way with food donations and Covid measures – the favela even had its own ambulance and local marshals to advise residents on preventive measures. Other favela groups had adopted some of Paraisópolis's organisational strategies. Now Renato wanted to follow Gilson's example and become a social entrepreneur, helping people of the favela he grew up in and making a career for himself out of it.

A charismatic, ebullient and creative man, Renato also had a sense of discipline about him. His grandfather fought in the Second World War for the Brazilian army in Italy and believed that only a military connection could save his family from falling into the criminal temptations of the neighbourhood. His father was a maintenance manager at an airline, which enabled them to travel. Renato studied biomedicine and became a salesman, and at the same time a musician who ran parties in the Jurunas favela where he grew up. Members of the gang that run the favela would come to them to sell drugs.

'It would have been easy for me to become a guy totally involved in this kind of marginal business – gangs, drug dealing. When I started doing parties, it was basically impossible not to have some kind of involvement because they were people who played football with us, jumped in the river with us,' he said. 'At the very least, we had to have a relationship with them.'

Music ran in his family. Renato plays traditional styles, like the swaying carimbó of Pará and modern pop and hip hop, doing shows at weddings and cultural events. He knew Gaby Amarantos, the Brazilian popstar from Jurunas who popularised the region's technobraga sound and now combines singing with being a judge on a TV talent show and acting in soap operas. Renato co-wrote a song on one album.

Back in 2019, Renato was teaching teenagers and young people how to DJ and rap through his project Farofa Black when he met Gilson Rodrigues. A month later, he set up OCAS. 'He showed me another vision, that this could become a big social business, and that the big companies basically have an obligation to help a project like this. What I saw was the sustainability of our biodiversity,' he explained. 'The main problem is sanitation.'

Now his plan was to hold parties playing traditional music because he believed that Amazon residents needed to revive their traditional ways, from when they lived within the forest and rivers without destroying or polluting them. As a young child, he used to

Stop Destructive Development: Managed Urbanisation

swim and play in the river behind us, until a cholera outbreak in 1991. 'What brings us hope that the forest can continue alive is the empowerment of our communities with the traditional way of life. We can't lose the standards, the forms, the processes which made the forest the sustenance of these peoples,' he said.

Renato knew about the poverty, the temptation of crime and how it impacted households like the two we visited. He talked about how common teenage prostitution was, how anything was on sale in a place with no laws. What the community needed was education and a chance, he argued. He had been to a riverside community in a nearby state that planned to make soap out of coconuts (coconut soap is widely sold in Brazil). He has since helped set up a project by young women who use an app to find river transport to a nearby island. He had his plan to produce bricks out of açaí stones and had found a location for a community allotment. He was full of grand plans.

'I didn't know it was possible to strengthen my name to the point where I could be someone who became a reference to receive funds capable of creating social projects,' Renato said. 'Because art has the power to turn the key in your head.' Since we met, he had laid out his vision in an Instagram post. 'One day we dream of being a platform to generate projects of social impact and be able to change the lives of people through art,' he wrote. 'We perceive that our opportunity to cause social impact is infinite.'

Thinking back to Jurunas, and people like Renato and Ana Claudia, that seems so clear. If the future of the Amazon is left to politicians, it is probably doomed: as in many countries, far too many politicians are more concerned with increasing their own considerable wealth and power than benefiting their country, its people or its nature. The long term is not a concept they are familiar with in the grab for more cash and contacts. Increasingly on this journey, I came across people with bigger visions, and they were invariably connected to their communities in a whole different way. They didn't live behind bars. They lived within them.

Dom reached this far in drafting his book by the time of his murder on 5 June 2022.

From this point on, his friends have worked with his plans and notes to complete the missing chapters.

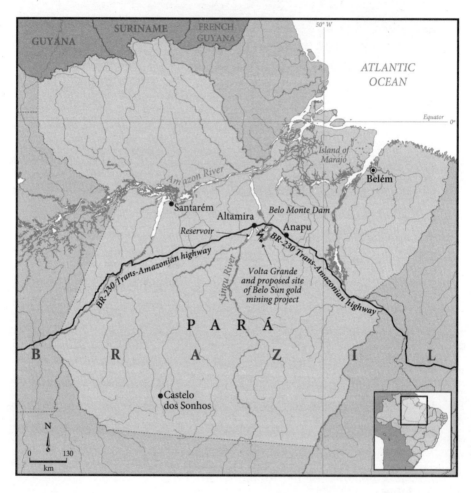

Selected infrastructure of Pará State

CHAPTER FIVE

A CEMETERY OF TREES: INFRASTRUCTURE CATASTROPHES

Eliane Brum and Dom Phillips
Altamira, State of Pará
(Translated by Julia Sanches and Diane Whitty)

Eliane Brum is a writer, journalist and documentary filmmaker based in the Amazon. She is the creator, co-founder and director of the trilingual journalism platform Sumaúma – Journalism from the Center of the World, a columnist for the Spanish newspaper El País *and one of Brazil's most award-winning reporters. In 2022, she received a Maria Moors Cabot Prize from Colombia University for her body of work. Three of her eight books have been translated into English:* Banzeiro Òkòtó – The Amazon as the Center of the World *(Indigo/Graywolf),* The Collector of Leftover Souls *(Granta/Graywolf) and* One Two *(Amazon Crossing).*

'They're killing us off little by little. Pretty soon there won't be anyone left to make demands.'

I remember pale blue eyes set in a face stitched by the sun. It was August 2021, summer in the Amazon, meaning 40 degrees Celsius and slim chances of clouds dotting the sky. I was at a protest led by land rights campaigner Erasmo Theofilo outside the courthouse in the city of Altamira, in the state of Pará. The protest was another chapter in the battle to establish a settlement in Anapu, the municipality that had swallowed up the body of Dorothy Stang, the US missionary who was assassinated in 2006 for fighting to secure land for small farmers engaged in agroecological projects. Dom and I shared a sliver of shade by a wall across the street.

We'd first met three years earlier, in 2018, on the beachfront in Rio de Janeiro. He, his partner Alessandra, Jonathan Watts and I met up for a drink. I was already living in Altamira, and Dom asked me why I kept saying the Amazon needed to be decolonised. At the time, the forest seemed far away from Dom, and it would have been hard to imagine finding him one day on a journey for answers to save the largest tropical forest on the planet. Later, he'd attend my wedding to Jonathan in London as a friend of the groom.

But suddenly, there we were, in one of Brazil's most violent cities, both staring out at the same scene. Dom asked if he could talk to me for his new book and I said I could meet him the next day. But he would be gone by then. This is how we came to never see each other again. I remember thinking his eyes were the same colour as the punishing sky in the Amazonian summer, possibly the only thing connecting that well-intentioned man from the northwest of England to the brutality unfolding beneath the vast blue. How could I have known that those same eyes would be gunned down in a different Amazon, closing their blue for all time?

I can't step into Dom's body to take up this writing project, born from a vision that was his alone. So I'll have to do it with my own eyes, which are the colour of tree bark or nut skins or even

A Cemetery of Trees: Infrastructure Catastrophes

of the earth where I've seen so many bullet-ridden bodies buried. I know what he saw. I've conducted countless interviews with the same people he spoke to and written dozen of articles, including a book of my own, about the Amazon. But what I see and hear may be different – and it's important for readers to know this.

If Dom and I had met up to discuss his book, I might have told him that the Amazon isn't so much saved as defended every day by people like Erasmo Theofilo, Raimunda Gomes da Silva, Antonia Melo, Juma Xipaia. I could have said that it's not us who save the forest but the forest who saves us, every day. And I don't mean that imaginary entity packaged as a large green mass. I mean the loud chatter of interlinked and interdependent beings of all shapes and sizes, both below and above ground. I may have told Dom that the forest is not about individuals or groups but about relationships. I might have suggested he use a different title for the book.

But we didn't meet up. And so it's fallen to me to take up the pen and write this chapter, not long after that brutal interruption of bullets, which we did not so much hear as feel.

Raimunda Gomes da Silva, whom Dom visited, has a lot to say about writing and pens. When she was ten years old, she started working as a modern-day domestic slave for a rich white family. The first thing she learned was to be soft-footed, a lesson taught to her not by her bosses but by her father. Black girls were to walk through the world without making noise, without disturbing the mistress of the house, without calling attention to themselves. To this day, Raimunda walks as if on air. And yet her father must have neglected to teach her to keep her mouth shut, and so she makes a huge racket the world over denouncing the Belo Monte hydroelectric dam complex. Her voice is a little shrill and she speaks fastveryfastswallowingsyllableshereandthere. Raimunda claims pens destroy – and she is absolutely right.

It's because of the wielding of pens that the Amazon is hurtling toward a point of no return. Although the onslaught on the Brazilian rainforest began much earlier, it only became an ideology by the pen of the military dictatorship that suppressed the country for 21 years, from 1964 to 1985. In the 1970s, the dictatorship developed a plan and a mythical social image around the Amazon that persists to this day. 'A forest without men for men without land' was one of the slogans broadcast in newspapers, television programmes and pre-show propaganda at movie theatres of that era. 'Green desert' was another. Sometimes also 'human desert'. The forest was a body to be dominated, violated, exploited and depleted. A woman's body, from the perspective of patriarchal rule-makers. In the eyes of generals, the Indigenous peoples, who'd not only lived there for at least 13,000 years but also, as has been demonstrated, planted a portion of the rainforest, were less than human, simply another obstacle to be cleared. Nothing could stop the 'march of progress' – razing the forest for highways, damming rivers for the construction of power plants, tearing ore from the bowels of the earth for exportation, planting vast swathes of monocultures like soybeans, creating pastures for thousands of cattle, founding cities over the ruins of the rainforest. The entire country was asked to praise and take part in the megaproject that would mark the supremacy of that dictatorship.

'On these banks of the Xingu River, in the middle of the Amazon jungle, the President of the Republic has set in motion

A Cemetery of Trees: Infrastructure Catastrophes

the construction of the Trans-Amazonian highway, a historical launchpad toward the conquest of this gigantic green world.' These are the exact words on the plaque unveiled by dictator Emílio Garrastazu Médici, commemorating the ground-breaking ceremony for of the Trans-Amazonian highway in the municipality of Altamira, on 9 October 1970, a road planned to cut through 3,000km of forest. A 50-metre-tall Brazil nut tree was chopped down to mark the occasion, symbolising the conquest of the jungle. Médici oversaw the bloodiest period of the dictatorship, the one with the most kidnappings, torture and assassinations of opposition members. At least 8,000 Indigenous people were murdered, most of them in the Amazon. The site of the uniformed man's highly phallic inauguration has come to be known by a telltale name: *Pau do Presidente*. It was right there, in Dom's notebook, in his almost illegible chicken scratch handwriting: 'joke – the President's wood / cock.'

The Trans-Amazonian highway was imposed by the pen of generals, opening the way for chainsaws, diggers and firearms. The highway pierced the Amazon, trespassing on the forest and on the forest's human and non-human people. It first established itself with blood. This is the road of invisible corpses that Dom took out of Belém, the capital of Pará state. Dom described his plans for this chapter as follows: 'Infrastructure is always sold to the public as development but it masks conquest and destruction. The main beneficiaries are the rich in faraway cities.'

Dom's first stop was Anapu, where he met the Catholic missionaries Jane Dwyer and Katia Webster, colleagues of Dorothy Stang's and just as fiery. Following Dorothy's murder, the city went through a more peaceful period because the killing made world news and caught the state's attention. Shortly after, the failures of Brazil's democracy were laid bare. First through mass street protests in 2013, then through a series of events in 2016 that culminated in the impeachment of President Dilma Rousseff of the Workers' Party. Many in the Left still believe that Rousseff's impeachment was the product of a coup orchestrated by Congress with the support of Rousseff's vice president and immediate successor, Michel Temer, of the Brazilian Democratic Movement.

One year before Rousseff was ousted from her elected position, the bodies once again began to fall in Anapu. The reason for this is that the forest's predators are like sharks, except instead of smelling blood, they sniff out the weaknesses of a cornered democracy and determine that the government is no longer an obstacle. In her final years in office, Rousseff had developed close ties to predatory agribusiness, as had other Workers' Party members. When Temer took office, the audacity of public land theft known as *grilagem* returned with a vengeance almost overnight. Blood returned to Anapu, as the land grabbers were more emboldened than in Dorothy Stang's time.

Jane and Katia, two US-born missionaries who chose Brazil and the Amazon as the stage of their lives and struggle, are rooted in a time when the Catholic Church still wielded power in the Amazon region. A significant portion of past and present forest leaders were formed in basic ecclesiastical communities that subscribed to liberation theology. Throughout his long papacy, conservative John Paul II fought against this movement that first took root in Latin America in the 1960s and viewed the teachings of Jesus Christ through the lens of the socioeconomic liberation of the oppressed. Joseph Ratzinger, an ultra-conservative cardinal who became pope in 2005 and stepped down in 2013, was tasked with issuing warnings to the movement's advocates, which in practice was an attempt to destroy it. This decision, at least in Brazil, partly explains the growing influence neo-Pentecostal Evangelical churches have on both our customs and our politics – more so in the Amazon than anywhere else in the country.

It is thanks to the work of missionaries like Jane and Katia that the Catholic Church still has a stronghold in Anapu. The main champions of liberation theology, as well as the fight for the forest and its people, have either died or been forced into retirement, as is the case most notably with Bishop Erwin Kräutler, who was Prelate of Xingu for over three decades, at least one of which he spent under police protection after receiving death threats. Forcibly retired, Kräutler is now 'prelate emeritus of Xingu'. Here, the word 'emeritus' signifies that the bishop in question no longer holds any power. The movement has been kept alive predominantly by nuns and missionaries like Jane and Katia who, being women, have received almost no support from the Vatican, even under the progressive papacy of Pope Francis.

Every 12 February, Jane and Katia help the community organise a political–religious ritual marking the 'martyrdom of Dorothy Stang'. Its main purpose is to keep the struggle for the forest and land reform going. Besides the bishop, there are often a dozen priests at the mass held in honour of the late missionary. Though men in cassocks occupy the stage, Jane and Katia are the ones running the show from the wings.

In the bloody years between 2015 and 2019, during which time the Pastoral Land Commission recorded 19 murders in that region, it wasn't uncommon for those praying at Dorothy Stang's cross on 12 February to become another name on that same list the following year. Now, the names of Bruno Pereira and Dom Phillips are also brought up during mass. How could Dom have imagined that, a mere year after visiting Dorothy's gravesite in 2021, he would join her as another martyr of the Amazon? And yet there he is, peering out at us from a poster on the wall by the shrine.

The Amazon's political–religious impasse is that the majority of nuns and missionaries sustaining its Catholic communities are now in their seventies, like Katia, or even their eighties, like Jane. And there are no successors in sight. The forest is becoming increasingly Evangelical, with most people following neo-Pentecostalism. Centred on the market and profit, this strand has ties to the predatory exploitation of the rainforest. Religion and politics are closely entwined in the Amazon – and all across Brazil.

One example: in February 2024, the world was shocked to find out that the man who murdered Chico Mendes (1944–1988) – environmentalist, trade union leader, rubber tapper and the most important martyr of the forest in recent history – had become an Evangelical pastor and president of the Liberal Party, of which Jair Bolsonaro is a member, in the municipality of Medicilândia. On the Trans-Amazonian highway, 85 kilometres from Altamira, in a city named after the bloodiest general of the Brazilian military dictatorship, Darci Alves Pereira invoked God and founded churches while preaching far-right ideas.

The people who kill today around the Trans-Amazonian highway are carrying on the dictatorship's ideology towards the rainforest. Large public works were a defining feature of the military government's infrastructure 'development' plan. Decades after the transition to democracy, these megaprojects were taken up again under the aegis of the Workers' Party and their Growth Acceleration Programme. Yet, in the 1970s, the Trans-Amazonian highway was the project that best symbolised the dictatorship's attitude toward nature. Back then, the region was split into two poles called 'Trans-West' and 'Trans-East.' The first section – the part marked for official colonisation and agricultural production – stretches from Altamira to Placas and mostly received settlers from the south of Brazil. The Trans-East region, on the other hand, which spans Altamira to Marabá, became the site of spontaneous colonisation, the kind consistently forgotten about in official public programmes. The majority of migrants to this region were from the Brazilian northeast and did not have official state support to occupy land deemed to be less 'productive'. Though let's not forget that for millennia, all this land, from the east to the west, was home to Indigenous peoples.

The Trans-Amazonian highway and 'the conquest and colonisation of this gigantic green world' began with the genocide of Indigenous peoples and continued with the renewal of policies for the racial whitening of Brazil's population. The first chapter of this project occurred during the Imperial period. It bears remembering that the south of Brazil was re-colonised in the nineteenth and twentieth centuries over the bodies of Indigenous peoples by immigrants brought over from Europe, especially Germany and Italy. Not only were Indigenous peoples driven out of their land and largely exterminated, but when the time came to decide who should replace them, the government chose to import white Europeans.

Brazil was the last country in the Americas to formally abolish slavery, in the late 1800s, after four centuries serving as the stage for the largest slave trade in the world. The abolition of slavery, however, did not come with plans for social inclusion. After centuries of trafficking the bodies of millions of enslaved Black Africans, what concerned the elite was ensuring that Brazil's population became whiter. The first public inclusion policy directed at the country's Black population was only enacted this century. Although the process began with Fernando Henrique Cardoso of the Brazilian Social Democracy Party, Brazil's president from 1995 to 2002, and was finalised while the Workers' Party was in office, it is actually the result of enormous pressure from Black movements that have gained prominence in recent decades.

During the construction of the Trans-Amazonian highway in the Trans-West, new colonisers were summoned from the south of Brazil, the majority of them descended from Europeans who had immigrated to the country in the previous century, their principal ambition being private land ownership. Things weren't easy for the European immigrants in the late nineteenth century, and nor were they easy for their descendants who followed the Trans-Amazonian highway in the 1970s. It soon became clear to them that they would have less support than the dictatorship had promised, in a strange forest that called for a different way of life, among Indigenous people fighting for their survival. Yet things were a lot harder for the men and women who arrived uninvited from northeast Brazil with no government support, chasing the dream of a patch of land, mostly in the Trans-East, to call their own, freeing them from the need to rent their bodies to large landowners.

In that same region of the Trans-Amazonian highway, the dictatorship implemented a land concentration policy whereby it preferentially awarded provisional titles to 3,000-hectare plots to

Dom *(left)* at home with brother Gareth and sister Sian. Bebington, 1972.

In the offices of Bristol's New City Press, which Dom co-founded in 1988.

Dom and his wife Alessandra Sampaio on his last birthday. Bahia, 2021.

Dom on the road in Yanomami Indigenous territory with his ever-present notebook, 2019.

The Amazon is the world's largest rainforest and home to one in ten of all species on earth.

Throughout much of the region, life takes place by and on the water. Marajó Island.

The Amazon is home to more than 350 ethnic groups: At the foot of a majestic samaúma tree, the families of Wewito Piyãko, Francisco Piyãko and Moises Piyãko, leaders and founders of the Ashaninka Apiwtxa village.

© Nicoló Lanfranchi

Huge infrastructure and mining projects are motors of environmental and social devastation in the Amazon. The Belo Monte dam, Altamira, 2019.

Cattle grazing on a ranch inside the Jamanxim Forest protected reserve. Cattle ranching, which provides beef for global markets, is by far the biggest driver of deforestation.

A rare photo of Dom and Bruno together. Tabatinga, Amazonas, en route to the Javari Valley, 2018.

Dom questioning Brazil's far-right president Jair Bolsonaro at the presidential palace in Brasília, 2019.

Bruno in the Javari Valley, 2018.

The FUNAI expedition led by Bruno Pereira (top right) which Dom (in red) accompanied through the Javari.

The disappearance of Dom and Bruno sparked protests around the world, first to urge Brazil's authorities to step up the search, and then, after the murders were confirmed, to demand justice. São Paulo, Brazil.

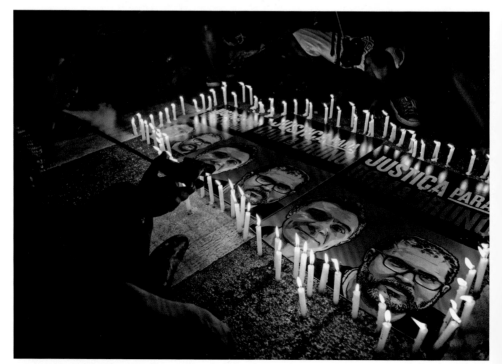

A Cemetery of Trees: Infrastructure Catastrophes

people from outside the Amazon. Oftentimes, these agreements were financed by the Superintendency for the Development of the Amazon, or SUDAM, an acronym that would become infamous for its part in corruption scandals. In order to receive land titles, candidates were given five years to demonstrate that an agricultural company had been installed on the land in question. Many of these plots were sold to third parties before even being granted the definitive title. In many cases, the government never took action.

Public land and public funding have produced but also stoked a speculative agrarian market in the Amazon, as well as a cycle of *grilagem* (land grabbing) and assassinations that persists to this day. All of this is closely related to large infrastructure projects. The word *grilagem*, which is essential to understanding the destruction of the Amazon, derives from the trick used to forge land deeds: they'd put crickets in a box with a false title so the insects would age the paper by yellowing it and riddling it with tiny holes. With the complicity of the well-remunerated owners of public notary offices, a caste that controls Brazilian bureaucracy, from birth, marriage and death certificates to real estate titles, the fraudulent document became legal. Up until the 1980s, public notary offices were passed down from father to son, following the same logic as dynastic monarchies.

Once again, the destructive pen Raimunda Gomes da Silva referred to. Even when false, written documents were worth more than the true oral testimony of Indigenous people who have lived in the rainforest for thousands of years; of *quilombolas*, descendants of enslaved African rebels whose ancestors have been living in certain regions of the Amazon for the past four centuries; and of *beiradeiros*, members of traditional forest communities who have lived in close relationship with the Amazon's rivers for more than a century.

The city that buried Dorothy Stang has been marked by carnage wrought by the criminals who perpetrate *grilagem*, often with the complicity of co-opted police officers. The dispute is between those who appropriate vast swathes of public land for speculative purposes and small farmers who wish to live off the land and grow food for the rest of the country. Because there is a power imbalance, this dispute has resulted in an ongoing massacre.

In 2018, a kill list was being passed around Anapu as casually as a list of back-to-school supplies. 'My son asked the head gunman if he was on the list and he said he wasn't,' a perplexed farmer told me mere days after her son was executed in their own home, the third in their family to meet that fate. Shortly after, someone knocked on their door with the following message: 'I'm here to tell you that two more of your family members will die.' They packed up and left. The landscape Dom Phillips encountered less than a year before he himself was gunned down is a direct result of the Trans-Amazonian highway and the ideology that inflicted it on the forest.

First *grileiros* (land grabbers) raze the forest and sell the wood. Then they bring in a few cattle to occupy the land, convince others that it belongs to an owner and fake its productiveness. Should they fail to legalise this theft through collusion with public notary offices or other expediencies, they can always count on Congress to put forward a bill that, under the guise of 'land tenure regularisation', will eventually grant them definitive ownership over that piece of land. Since the turn of the century, two such bills have been passed: one in President Luiz Inácio Lula da Silva's second

A Cemetery of Trees: Infrastructure Catastrophes

term and another under Michel Temer. This is how *grileiros* go from thieves of public lands to legal landowners. Obviously, there is no better business than appropriating land and selling it: barring the cost of chainsaws, contract killings and semi-indentured labour (which is by definition cheap), it's pure profit. Dorothy Stang was killed for helping smallholders legally remain on land coveted by *grileiros*. As was the case with other leaders whose executions raised barely a whisper because they were poor and had neither British nor US citizenship.

For forest peoples, development and infrastructure are predatory motions. In Portuguese, the prefix 'des' indicates that a thing has ceased to be or do something; it marks the opposite of what follows. (Much like the English prefix 'un'.) The Portuguese word for 'develop', *desenvolver* (or 'un-involve'), therefore can be understood to mean 'cease to be involved' – with the forest, with its people, with life, with the consequences of their actions, with the impact of infrastructure projects. Dorothy and forest leaders, both the fallen and those still living today, have reclaimed the idea of 'involvement' – to be involved with the forest, to connect, to become implicated, to see ourselves as interdependent. These are two radically different life projects. This is the conflict in the Amazon and in every biome: those who are involved die at the hands of the uninvolved.

When I met Dom at the smallholders' protest, he was following Erasmo Theofilo, one of Anapu's most important leaders. Erasmo, whose prefixes are they and their, is an extremely impressive human person, which is exactly how they self-define. Erasmo had polio as a child and can only get around in a wheelchair. Or, as is their preference, in an ordinary chair, the white plastic kind you find at cheap bars and that Erasmo incorporates into their own body, using it to almost run. Besides being a disabled person in a place where, unlike cities, it's impossible to ignore our own bodies, because our bodies are always being asked to show up, Erasmo is openly bisexual in a land full of armed 'machos'.

Erasmo had recently received death threats. In 2019, far-right president Jair Bolsonaro's first year in power, Erasmo and their *quilombola* partner Natalha were forced to leave Anapu around Christmas in the hopes Erasmo would have a chance of ringing in the new year. This is what life is like on the frontline of the war against nature. Over Christmas and New Year, institutions go on call and large environmental organisations close their offices, suspending the meagre protections afforded to people whose lives are in danger. So, at a time when most people are travelling to spend the holiday season with their family, and commercials encourage consumerism through gift-buying, lavish feasts, drinks and New Year's Eve outfits, in the forest, leaders are forced to leave behind their homes and families and land in order to stay alive. It's like this year after year. Holidays are a perilous, excruciating time for those who defend the Amazon with their bodies.

Every year, following attempts on their lives in their own home, Erasmo and Natalha have had to flee for longer stretches of time to keep themselves safe, taking Natalha's four children with them, who are forced to miss school. On 15 April 2021, Natalha and Erasmo's son Eduardo was born during one of these escapes, into exile in his own country. The family's hideout was discovered and hooded men attempted to kill them. Eduardo barely knows his land of origin as in 2022, Erasmo and Natalha had to leave Anapu forever after someone threatened to kill their children 'to touch Erasmo's heart'. Lula's return to power in 2023 had no effect whatsoever on the violence.

The massacre of the forest is the massacre of the human people who defend it. When the government fails – deliberately, in Bolsonaro's case – to protect those who protect the Amazon, leaders have no choice but to leave their land so as not to be martyred like Dorothy Stang and so many others. Deprived of this leadership, the resistance weakens and communities become more and more vulnerable. For those forced to leave their land and their struggle, this is just another kind of dying. Uprooted from the communities they belonged to, pushed out of the struggle that gave meaning to their lives, living alone in unfamiliar places, in invariably precarious conditions, the forest's most fervent protectors are the ones most punished, even when they are not executed. Some attempt suicide.

The predators of the forest understood this full well. These days, they know it makes less sense to murder, which could bring the state close and get in the way of their business, than it does to drive people out. On 10 January 2024, the little babassu-straw school the community built for its children was set on fire a second time. *Grileiros* target schools to drive out adults and, while they're at it, traumatise their sons and daughters, who grow up in fear. Some of these children have seen men holding guns to their parents' heads inside their own homes. Others cannot bring themselves to return to the school they saw in flames. This tactic, which is carefully considered and executed, strikes at the heart of the movements. It also jeopardises leadership succession by undermining new generations. All this within Brazil's tattered and ever-endangered democracy.

HOW TO SAVE THE AMAZON

People like Erasmo defend the forest for all humankind, for every reader of Dom Phillips' book. This cycle of threats, deaths and human rights violations is Erasmo's life because this same humanity's inaction is a kind of action. Dom had gone in search of 'solutions' to save the Amazon. In the case of the land tenure war in Anapu and various other parts of the Brazilian side of the tropical rainforest, any solution must involve land reform.

Some of Brazil's ecologists and environmental organisations view smallholders as forest 'invaders', a misunderstanding that has cost leaders their lives. Family farmers are in the Amazon to stay. Without support, they can end up working for *grileiros* or join the ranks of the organised crime factions swiftly advancing upon the forest. In many regions, smallholders are the only ones standing between the forest and its deforestation for the purposes of real estate speculation, soybean monocultures and pasture for cattle. Agrarian settlements must be created, recognised and supported so these families can stay on their land – provided they adhere to the principles of agroecology, refrain from using pesticides and exclusively produce food. Family farming is the eighth largest food producer in Brazil. But some of the men and women growing the food that feeds Brazil are being mowed down along with the forest.

The state, Raimunda Gomes da Silva reminds us, must cease using its pen to destroy and instead, in Anapu's case, use it to sign the documentation for land reform and other projects that will allow smallholders to remain on their land in collaboration with the forest. Schools must stop burning and guns must be kept away from children's heads. In fact, the sons and daughters of smallholders must be tended to so that they may take up their pens and write the future in safe classrooms inside a protected forest.

A Cemetery of Trees: Infrastructure Catastrophes

One hundred and thirty-seven kilometres of Trans-Amazonian highway and a ferry ride across a river separate Anapu from Altamira, a major hub in southwestern Pará. With a population of 126,000, Altamira is one of the most violent and deforested municipalities in the Brazilian Amazon. It is also the biggest municipality in Brazil and one of the biggest in the world, at 160,000 square kilometres, which is larger than half the world's countries. To give you a better sense of this Amazonian madness, Castelo dos Sonhos – 'Castle of Dreams' – a district of Altamira known for its frightening surge in violence, even by Amazonian standards, is approximately 1,000km from the municipal seat, along a road that's rarely in good condition, with not a single airport on the way. If public policies are precarious all over Brazil, and even more so in the Amazon, what chance do they have of making it to remote locations like Castelo dos Sonhos? I've told this district's story through its cemetery, where sometimes a victim lies beside their gunman, who was in turn killed by another gunman, also now planted a few feet underground. It is a cemetery where the gravedigger always has the forethought to leave a couple of graves open, knowing it won't be long before they are needed.

This country-sized municipality is one of the epicentres of the Amazon's destruction, oftentimes taking the prize for the highest rates of deforestation and arson. It is also at the centre of two infrastructure projects in the Amazon that are most symbolic of Brazilian government policies. Paradoxically, they were carried out during two very different political periods. While the Trans-Amazonian highway was built under the military dictatorship, the Belo Monte hydroelectric dam complex was constructed during the most left-wing governments in the history of Brazilian democracy:

the Workers' Party's 13 years in power, first under Lula (2003–2010) and then later under Dilma Rousseff (2011–2016). This paradox is proof of a phrase Tom Jobim once used to describe his country: 'Brazil is not for beginners.' Meanwhile, Altamira and Belo Monte confirm that Tom Jobim could have added: '. . . or experts.' Even academics and experienced journalists have failed to understand a country the size of a continent that holds within its geopolitical body such starkly different worlds.

The Amazon continues to be viewed as a body to be dominated, violated, exploited and depleted, even by the most left-leaning administrations. To the extent that the main environmental disputes during Lula's third presidential term were over a proposal to open another front of oil exploration in the Amazon, a project to build a railway to transport soy and other grain across the forest, and a plan to upgrade the BR319, a road through one of the last remaining intact areas of the forest between Porto Velho and Manaus. This, despite the fact that road construction is always a harbinger of greater destruction, as Dom wrote in his notes.

The year 2023, the first of Lula's third term, marked what a widely respected group of climate scientists called humanity's entrance 'into uncharted territory'. Temperatures hovered around 1.4 degrees Celsius above the pre-industrial era, the year was called 'the hottest twelve months in 125,000 years', and there was a record number of extreme weather events all over the planet. The rainforest went through the worst drought in the recent memory of its inhabitants, altering the ecosystem to such a degree that a village in the state of Amazonas vanished entirely, as if it had been swallowed up. That same year, Lula's government used the phrases 'Amazon rainforest' and 'fossil fuel exploration' in the same breath.

A Cemetery of Trees: Infrastructure Catastrophes

The Belo Monte hydroelectric power plant, then called Kararaô in the language of the Kayapó people, was conceived during the military dictatorship. The plan was to dam the Xingu River, which is a tributary of the Amazon and one of the most powerful in the rainforest. Yet the dictatorship was met with fierce resistance from Indigenous peoples and social movements in the affected region. The most symbolic image of this resistance occurred during a gathering of Indigenous peoples of the Xingu in Altamira in 1989, when Tuíre, an Indigenous Kayapó woman, touched her machete to the cheeks of the director of Eletronorte, then a state-run power company, and spoke the following words: 'Electricity will not give us food. We need our rivers to flow freely. Our future depends on it. We do not need your dam.'

Thanks to this resistance, the power plant plan was shelved for decades. In 2002, the year Lula won office as the great figurehead of the Workers' Party, of which several social movement leaders were members, including regional founders of the party, like Antonia Melo, there was widespread relief. Local supporters believed the Xingu River basin was finally safe from the spectre of the power plant, which haunted it with every new government, even after the country's transition to democracy. Campaigners let down their guard; meanwhile, plans were redrawn in order to ostensibly diminish the dam's impact on the rainforest's delicate ecosystem. The power plant went to auction under the name Belo Monte in 2010, the last year of Lula's second presidential term, in a deal with serious indications of fraud. A consortium of construction companies immediately got to work.

In 2015, this consortium would be accused of corruption by a federal investigation known as Operation Car Wash, which led to the imprisonment of several owners of large construction companies – along with Lula. For the first time in Brazil's history, an anti-corruption investigation had touched the cartel of construction companies that had a hand in major projects undertaken in the Amazon since the time of the military dictatorship, and some of those who grew rich from corruption schemes tied to large infrastructure projects were sent to prison. However, the alleged illegalities committed in the service of the law, including the controversial, abusive participation of the Public Prosecutor's Office and former federal judge Sergio Moro, who left his court position in 2018 to become Jair Bolsonaro's minister of justice, jeopardised the many victories of Operation Car Wash.

Yet, by then, Belo Monte had already been built. The dam forever altered the dynamics of the Xingu River and its tributaries, driving out hundreds of families of *ribeirinhos* – members of traditional forest communities who lived on the islands and riversides that were flooded – and radically changing the way of life of Indigenous peoples, some of them, like the Arara people of Cachoeira Seca, only recently contacted.

With little alternative, large numbers of these displaced persons were forced into the city of Altamira, almost doubling its population. Many were rehoused in peripheral communities that became fertile territory for drug gangs. Altamira quickly became the most violent large city (of more than 100,000 residents) in the country.

The expansion of organised crime in the region accelerated. On 29 July 2019, the city suffered the second largest prison massacre in Brazilian history, leaving 58 dead. Another four incarcerated people were executed during their transfer to Marabá on the Trans-Amazonian highway, bringing the total number of victims to 62. Desperate mothers, wives, sisters and daughters stood outside the morgue, many holding children by their hands or in their arms, wanting to know if their loved one had been decapitated or burned alive.

At the start of the following year, shortly before the first Covid-19 death in the city, Altamira was shaken by a series of suicides, mostly children and teenagers. Between January and April 2020, 15 people in Altamira killed themselves: nine of them were young, from 11 to 19 years old, one was 26, and the other five were between the ages of 32 and 78. Two teenagers threw themselves off a tower in the city but the majority died by hanging. The only reason this number was not higher is that a group of mothers of murdered teenagers – yet another tragedy that flared up with Belo Monte – got together to organise an emergency operation to identify suicide warning signs on social media. Some days, as many as three suicide preventions would be carried out by this rescue network, which was hurriedly put together without any help from the state.

Mental health specialists came to a unanimous conclusion: the main hypothesis for this epidemic of suicides was the transformation of the city and its residents' way of life by Belo Monte. The teenagers who were dying by suicide had been children back when the power plant drastically changed the fabric of their families and the city, eating away at the quality of their everyday relationships, leading to increased incidences of violence (including domestic violence), alcoholism and substance abuse, and illness (including mental illness), with no adequate response to a change of this magnitude by public authorities. Organised youth collectives sent a stinging letter to the powers that be: 'They say we are the future of this country, but how can we be the future if we have no present?'

The impact of Belo Monte on Indigenous peoples was devastating and its effects continue to this day. For approximately two years, during the period of transition, Indigenous communities received monthly rations of snacks, soda, sugar, instant noodles and other overprocessed foods, along with box-spring beds, motorised boats and flat-screen TVs. Some Indigenous leaders were even given pickup trucks. Various communities stopped growing food and fundamentally changed their way of life.

These funds could have been put towards preventing and reducing the project's impacts. Instead, they were used to purchase our current millennium's version of trinkets and beads. According to data from a dossier on Belo Monte published in 2015 by the Socioenvironmental Institute with the collaboration of experts from various fields, these handouts triggered 'one of the most perverse processes of co-option of Indigenous leaders and social disruption fostered by Belo Monte'. The Special Indigenous Public Health District in Altamira, part of the Health Ministry, said in a document: 'Preceding construction of the Belo Monte Hydroelectric Power Plant, Indigenous people began receiving food assistance in the form of packages of nonperishable, processed groceries in September 2010. As a result, they stopped putting in crops and producing their own foodstuff. However, in September 2012, this so-called benefit was cut, and Indigenous people were left without this food supply and also without gardens they could harvest for sustenance, driving the incidence of children with Low or Very Low Weight-for-Age to 97 cases, or 14.3%.'

A Cemetery of Trees: Infrastructure Catastrophes

Elsewhere in the document, the agency responsible for Indigenous health drew a link between the 2010 spike in cases of 'acute diarrheal disease' and activities in the villages, by the hydroelectric concessionaire Norte Energia: 'We recorded a considerable increase in 2010, with 878 cases reported in a population of 557 children under the age of five, equivalent to 157 per cent of this population, or 1,576.3 per 1,000 children. [. . .] Changes in dietary habits following the introduction of processed food—financed as part of the environmental licensing requirements for construction of the Belo Monte hydropower plant—likewise contributed to the high rate.' According to data from the dossier, malnutrition in children surged 127 per cent from 2010 to 2012, and a quarter of the children surveyed were malnourished. In the same period, healthcare services provided to the Indigenous soared 2,000 per cent in towns within Belo Monte's radius of impact.

The measures taken by the company, which set off a chain reaction, were part of the official mitigation plan for the damages caused by the dam – making it all the more twisted.

Yet the Rousseff administration granted the hydropower plant an operating licence in November 2015. Belo Monte had been built 'through a jumbling of the developer (the concessionaire Norte Energia) and the State', in the words of Federal Prosecutor Thais Santi, who filed suit for Indigenous ethnocide. Some of the measures upon which the project's licence was contingent have yet to be put in place.

This was how Raimunda Gomes da Silva came to know the weight of the pen that destroys. She was living with her husband João on one of dozens of islands on the Xingu. To fully understand these geopolitics, readers must realise that the Amazon's river islands are the last refuge for forest peoples. Just as the ancestors of today's Indigenous people once fled the colonisers' boots and guns by venturing ever deeper into the woods, traditional populations have fled encroaching land grabbing in recent decades by leaving the riversides and making their homes on islands where they thought they would be safe. Belo Monte proved they wouldn't be.

Otávio das Chagas was on his island with his wife Maria and their nine children when officials from Norte Energia, the consortium that owns the Belo Monte concession, arrived on the scene to announce the end of his world. They told him if he didn't sign the papers and leave the place where he had spent his life, he would be evicted without any compensation. Like many *ribeirinhos*, Otávio was only schooled in rivers. He knew nothing of the alphabet. Coerced and afraid, the fisherman, who stands less than five feet tall, used his thumb to sign a document written in lawyer's words he couldn't decipher. And this is how, in 2014, Otávio suddenly found himself on the poor outskirts of Altamira. As compensation for the loss of his island, his way of life, and his ability to fish and plant, they paid him 12,000 reais – less than US$5,000 in today's terms. The bones of Otávio das Chagas's father now lie crushed beneath the Belo Monte dam.

Hundreds of families were, like Otávio, cast far from the river by the might of the pen, moved to subdivisions that were planned by the dam company and immediately overrun by various factions of organised crime. The houses were cheaply built and their design incompatible with these people's way of life. Cut off from their community, they watched everything they knew disintegrate. Many got sick, some died.

A Cemetery of Trees: Infrastructure Catastrophes

Raimunda Gomes da Silva witnessed her husband, João Pereira da Silva, become, in her words, a 'living dead'. He went to the Norte Energia offices to 'negotiate' a settlement for the house he was losing. Wielded by the lords of the dam, the word 'negotiate' represented yet another moment of indignity, since there is no negotiation when one side has no choice. João told me that when they informed him of the pittance he would receive as compensation for everything he had managed to build for himself, he realised he was going to starve to death. He didn't have the strength to start over at the age of 63. That's when he decided to kill a company official. 'I would sacrifice my own life but draw the world's attention to what they were doing to people.' But João isn't a murderer. And so João's voice locked up, his legs locked up, and he had to be carried out of the office. Later the doctors would tell him he'd had a stroke. And then he would have another.

With little mobility in his legs, João said over and over that all he saw was 'darkness', while Raimunda travelled up the Xingu to their island to try to salvage their belongings. Well before she got there, she saw smoke rising into the sky. By the time she reached her patch of ground on the Xingu, it had all burned down. Amid the ashes, Raimunda sang to her plants, asking them to forgive her for failing to save them from the fire. When she got back to town, she reported the crime to the Federal Police. Then João asked Raimunda to go with him and their daughters to the incinerated island so they could all kill themselves there. He said: 'This way, the world will know Belo Monte killed us.'

When Dom Phillips visited Raimunda, she had already settled near the dam reservoir, where the river is dead and fish don't thrive, in a patch of forest she calls the Promised Land. She had managed to secure a little corner of the world for herself after fighting hard alongside other *ribeirinhos* for the chance to take up her pen and write lines of justice at least once. But she remains one of the few who have been resettled in the forest and she feels under constant threat. In early 2022, I accompanied her to the police station to report a death threat she had received while picking up her social security payment at a bank branch in Altamira. Raimunda is on the Ribeirinho Council, which was created by affected families with the support of the Federal Public Prosecutor's Office and the Brazilian Society for the Advancement of Science to demand that the hydro plant company establish a Ribeirinho Territory next to the reservoir so this traditional population can return to their way of life.

In the boat to Raimunda's house, Dom was horrified at the sight of the trees drowned by the reservoir of Belo Monte – groves of corpses standing in the reservoir waters, all the way along the river there. He wrote: 'A forest of dead white trees, stark branches pointing accusingly at the sky.' Dom then asked the captain of the motorised canoe to head closer to one of the groves and the man replied: 'We are going to see a lot more further on. It is all because of the dam.'

Dom's observation, in his notes for the chapter he never wrote: 'Belo Monte is a mass graveyard of trees, their trunks serving as their own tombstones. An endless parade of dead wood, killed so that people thousands of miles away could have electricity cheaper than they pay for it here. That's the deadly cost of Brazil's Amazon development—an endless graveyard of dead trees.'

Just as the scientists warned before the hydro plant was even built, Belo Monte produces far less energy than its installed capacity because the Xingu runs very low in the dry season that is the Amazon summer. Often, only one or two of the plant's 18 turbines are in operation during dry months.

After receiving its operating licence, Belo Monte set off a process of ecocide in the Volta Grande do Xingu, a 130-kilometre-long, biodiverse bend of the river, home to three Indigenous peoples and several *ribeirinho* communities. By seizing 70 per cent of the river's water for its turbines, Belo Monte is drying up the Volta Grande do Xingu. Year after year, its existence hinders spawning and deprives the forest peoples of fish, the basis of their diet.

In 2024, despite the efforts of the Federal Public Prosecutor's Office and the fight waged by impacted communities, an irresponsible water management programme continued killing off the Volta Grande do Xingu. And the Ribeirinho Territory was far from becoming a reality. 'It took them five years to build the dam and we've been waiting eight years for them to create the Ribeirinho Territory,' said Maria Francineide Ferreira dos Santos. A *ribeirinha* and Evangelical Christian forged in the struggle for justice, Francineide thinks of the delay in these terms: 'They're killing us off little by little. Pretty soon, there won't be anyone left to make demands.' Every year, some of those who wait fall sick, while others give up. A number have died waiting.

Raimundo Berro Grosso is one of the *ribeirinhos* who had a stroke while waiting for justice. Expelled from the river and relocated to one of Norte Energia's planned subdivisions, he put a canoe in his living room, where it floats through dry air. Like all of these people, in the city he discovered what life is like when you have to pay to eat – and you don't have the money to pay to eat.

'Being rich means not needing money. And being poor is having no choice,' says Berro Grosso – 'Big Bellow', a nickname he earned from a voice whose strength was seized by suffering and disease. 'I was rich. Belo Monte made me poor.'

Raimundo Berro Grosso is wasting away but he knows what he's saying. Major construction projects, megaprojects like Belo Monte, are machines that convert self-sufficient forest peoples into the urban poor. When they lose physical territory, these people also lose their identity. And with it, their rights. They cease to be Indigenous, *ribeirinhos*, *quilombolas* – people whose rights are expressly recognised in the Brazilian Constitution. Instead, they are transformed into a generic category called 'poor', living on city outskirts and in favelas, where they remain as remnants, ruins, like the forest. This stratagem is the most effective way of advancing into the Amazon because it eliminates the resistance.

Meanwhile, on part of the land they left behind in the Volta Grande, a Canadian mining company is trying to press ahead with plans to build a giant gold mine – Belo Sun – that would strip more of the land and pollute the diminished waters of the Xingu. Infrastructure projects beget more infrastructure projects – all designed to siphon natural wealth out of the region.

There is a play called *Altamira 2042* that depicts the sounds of people of the forest, both human and more-than-human, during their struggle against Belo Monte. The central voice belongs to Raimunda Gomes da Silva. At the end of the play, audience members are invited to destroy the dam with their hands or a sledgehammer – a scene of total catharsis.

Dom Phillips wanted to find solutions. Freeing the Xingu is the dream of everyone who has placed their body on the frontlines

A Cemetery of Trees: Infrastructure Catastrophes

in the fight for the Amazon. But until the river is freed, the government must move its pen to bring about justice and reparations, forcing Norte Energia to finish establishing the Ribeirinho Territory, to fulfil the commitments that never left the paper they were written on, to implement a water management system that allows the Volta Grande do Xingu to live on. The state itself – not just this or that government administration – must pledge not to allow any new hydroelectric power plants to be built in the Amazon. Hydropower in the Amazon isn't clean energy: it's energy stained with blood.

Because no bullet can silence Dom, his book has been finished. Though the last word, at least in this chapter, comes to us from Raimunda Gomes da Silva, who represents the forest. In late 2023, she took up a pen herself and published her own book, which came with a little bag of medicinal herbs picked in her Promised Land. She wrote: 'I thought about wanting to be an eagle so much that one day I dreamed I was flying. Flying high. I could see all the trees. But when I was about to descend, I didn't know how. I knew how to go up, but I didn't know how to come down. Then I had another idea: I wanted to go up, so why would I want to come down? I'm going to stay right here.'

Roraima State

CHAPTER SIX

REGROW AND SELF-PROTECT: INDIGENOUS DEFENDERS

Tom Phillips and Dom Phillips
The Yanomami and Raposa Serra do Sol territories, State of Roraima

Tom Phillips is The Guardian's *Latin America correspondent. During two decades as a foreign correspondent, he has reported from nearly 30 countries, including Brazil, China, El Salvador, Haiti, Indonesia, Mexico, Nicaragua, the Philippines, South Korea and Venezuela. He started his career in Rio de Janeiro and was* The Guardian's *correspondent in Beijing and Mexico City before returning to Brazil in 2020. Tom has worked extensively in the Amazon since he first travelled there 25 years ago, including in the Javari Valley region where Dom and Bruno were killed.*

'Ya temi Xoa!' *(I am still alive!)*

The Ye'kwana toddler gripped the back of Dom's seat and burst into tears as their single-engine Cessna swept into the sky over the city of Boa Vista and banked west towards the largest Indigenous territory in Brazil. But the boy's sobs quickly subsided as the sun-baked savannah surrounding the airport gave way to a thrilling vista of seemingly endless jungles and caramel-coloured rivers that wound through the woodland below like calligraphy. Dom and the child gazed out of the plane's starboard windows in silence, spellbound by the natural beauty of one of South America's least explored corners.

'A thousand shades of green . . . wild untouched forest,' Dom wrote in one of the burgundy notebooks he had taken on his latest mission to the Amazon, in November 2019. 'Its magnificence chills the stomach . . . Why would anyone want to destroy this? What could possibly justify such a monstrous act?'

The plane was on its way to Waikás, a village along the Uraricoera river in Brazil's Yanomami Indigenous territory, where about 150 members of the Ye'kwana people lived – among them Dom's anxious young flying-mate, Raimon, whose uncle had agreed to be his guide.

At the time of Dom's visit, Waikás was at the eye of a dangerous and fast-growing storm: a twenty-first-century gold rush that had been gaining momentum ever since the far-right populist Jair Bolsonaro became president almost 12 months earlier, in January 2019. On the eve of Bolsonaro's inauguration, Dom had spoken to the legendary Yanomami leader and shaman Davi Kopenawa, who voiced deep concern over the future of his isolated territory, which is home to about 30,0000 members of the Yanomami and Ye'kwana peoples, including some with little or no contact with the outside world.

'They don't want to respect where we live,' Kopenawa warned of Brazil's incoming right-wing administration. 'The forest is a sacred place for the Yanomami people . . . we don't want the whites to ruin it.'

Nearly a year later, as Dom's plane took flight over Roraima, Brazil's northernmost state, that was precisely what was happening.

Regrow and Self-Protect: Indigenous Defenders

Thousands of wildcat miners called *garimpeiros* were invading Yanomami lands, encouraged by Bolsonaro's anti-environmental and anti-Indigenous rhetoric and policies, which included dismantling the protection agencies supposed to defend such areas. Funded by politically connected, multi-millionaire criminal bosses who lived in luxury homes, some of these impoverished miners used aluminium boats to sneak up rivers like the Uraricoera to hundreds of illegal mines. Others used planes and helicopters to reach isolated jungle pits where minerals such as gold and cassiterite were plundered from lands the government had vowed to protect when the Indigenous territory was created nearly 30 years earlier. Bolsonaro's government, for the most part, simply shrugged its shoulders or, in the case of the president, actively cheered the miners on.

The Bolsonaro-era gold rush was, in many ways, a reenactment of a historic scramble for minerals that devastated the same region's Indigenous communities when Kopenawa was a young man. The military dictatorship's decision to colonise the Brazilian Amazon during the 1970s, bulldozing highways through its jungles, sparked a chaotic and bloody race for land and resources which is playing out to this day. During the 1980s, soaring gold prices and a punishing economic crisis saw as many as a million wildcat miners fan out across the mineral-rich region hoping to find their fortunes in Brazil's jungle goldfields. 'The Amazonian rush actually outstrips the Klondike and California gold booms of the last [nineteenth] century, both in terms of production and the number of people employed,' the British academic Gordon MacMillan, who lived with Roraima's miners at the time, wrote in his 1995 chronicle of the period, *At the End of the Rainbow?*

HOW TO SAVE THE AMAZON

MacMillan called the Amazon gold boom of the 1980s 'a major event in global mining history'. For the Yanomami – whose lands were located above some of the most coveted deposits – it was a disaster. By MacMillan's reckoning, at least 15 per cent of the Yanomami's relatively small population was wiped out between 1988 and 1990 by diseases inadvertently imported by nomadic bands of miners. The Indigenous NGO Survival International estimated that a fifth of the Yanomami died in just seven years. 'They are earth eaters covered with epidemic fumes,' Kopenawa would later write in his memoir.

The crisis – and the indefatigable activism of figures such as Kopenawa and campaign groups like the Pro-Yanomami Commission – sparked a global outcry. Public figures such as the Prince of Wales, now King Charles of the United Kingdom, took up the Yanomami cause. 'The Yanomami of Brazil are being driven to extinction by measles, venereal disease and mercury poisoning following the illegal invasion of their lands by gold prospectors,' the future king warned in a 1990 speech about how to save the world's rainforests from 'total destruction'.

Facing increasing international pressure and protest, Brazil's government decided to act. Thousands of miners were expelled from the region during a major security operation, which saw illegal airstrips dynamited and *garimpeiros* evacuated in military cargo planes. In 1992, exactly 500 years after European colonisers first reached South America, with dire consequences for its Indigenous populations, the Yanomami were granted a 9.6-million-hectare protected territory, a move Bolsonaro – then a fringe politician known for his crackpot ideas and bellicose style – unsuccessfully opposed. But nearly three decades later, Kopenawa and many

Regrow and Self-Protect: Indigenous Defenders

others feared history was repeating itself under Bolsonaro, who had boasted of working as a *garimpeiro* before entering Congress in the early 1990s and was notorious for his hostility to Indigenous rights. During a 1998 speech, the politician celebrated the 'competence' of the cavalrymen who had 'decimated' Indigenous populations in North America while lamenting that the same had not happened in Brazil. The unmissable implication was that, in Bolsonaro's view, there might not have been any Yanomami people left standing in the way of Brazil's economic development had its troops done their job 'better'.

As Dom flew into the Yanomami's Portugal-sized domain with his friend the Brazilian photographer João Laet, he had a dual mission.

Most urgently, he hoped to document and expose the criminal race for resources that Bolsonaro's shock election had unleashed on Indigenous territories across the Amazon. Activists from Hutukara, the Indigenous association Kopenawa helped found in 2004, had invited the journalists to make a rare visit in order to chronicle the devastation in *The Guardian* and, perhaps, embarrass Bolsonaro's government into taking action.

During their eight-day reporting trip, Dom and João planned to visit some of the worst-hit corners of the territory and talk to both the Indigenous leaders and illegal miners on the frontline of the growing crisis. 'We knew there was this big, ugly shadow hanging overhead,' João told me later.

But Dom was also travelling there in search of potential solutions that he could include in this book. How, he wondered, could Indigenous territories such as this one repel an increasingly hi-tech onslaught from illegal miners, loggers, ranchers, poachers, land grabbers and missionaries?

How might Indigenous lives, cultures, languages and lands be protected from the massive influx of outsiders who were egged on by government policies and the populist rhetoric of politicians such as Bolsonaro?

How could their rainforest home be saved from seemingly inevitable destruction? Or was the Yanomami territory, and indeed the Amazon itself – nearly 20 per cent of which had been lost in the last 50 years – already doomed?

Waikás, Dom hoped, might offer some clues. Not least because it was one of several villages in the Yanomami territory involved in a fledgling cacao planting scheme supported by the Brazilian NGO Instituto Socioambiental that was designed to generate income for such communities by selling the fruit to organic chocolate makers in wealthier parts of Brazil. 'We're trying to build another possible future,' Moreno Saraiva, an anthropologist involved in the initiative, told Dom.

The cacao bean project was part of a broader movement to find sustainable ways of making money from the immense biodiversity of the Amazon – thought to be home to 14,000 to 50,000 different species of plant – so as to ensure its remaining forests were worth more left standing than being torn down. A few weeks before his murder, Dom contacted one of Brazil's top biodiversity advocates, the climate expert Carlos Nobre, in the hope of discussing just that. 'I do think it's possible to save the Amazon,' Nobre insisted when I called him to conduct the interview about that movement which Dom had been prevented from doing.

Regrow and Self-Protect: Indigenous Defenders

Activists such as Nobre and those behind the Yanomami cacao project hoped that, if successful, such schemes might provide a blueprint for how Indigenous and traditional communities in the Amazon could shield themselves from decades of destructive and predatory extractivism.

Such aspirations would have been music to Dom's ears as he travelled the Amazon in search of ideas about how it might be saved. After visiting Waikás, he described the skinny cacao saplings he had seen there as 'green shoots of hope in a land scarred by the violence, pollution and destruction wrought by illegal gold prospecting'.

But as Dom made the nerve-jangling one-hour flight from Boa Vista to Waikás's hair-raising grassy runway ('From the air seemed much too small,' he wrote in his diary), it was the immediate threat to Indigenous communities that would have been foremost in his mind. All over the Amazon, the descendants of Brazil's original inhabitants were fretting over how the rise to power of a dictatorship-admiring conservative such as Bolsonaro might affect their lives.

The brutal nature of the rush for resources prompted by Bolsonaro's success was clear within hours of Dom's arrival in Waikás, which is located at the heart of one of the parts of the Yanomami territory worst affected by mining. Outside the village's health post, he and João came across two *garimpeiros* wrapping a man's corpse in a tartan hammock. The victim wore black football shorts with the emblem of São Paulo club Corinthians, his eyes were wide open and there was a white bandage over a bullet hole in his chest. The miners claimed, improbably, that the 21-year-old had taken his own life.

Giselle Dornellas, a young government nurse whose job was to care for the region's Indigenous residents, told the journalists she had spent hours trying to save him.

'She'd given him mouth-to-mouth. She'd pumped his heart. She put him on a drip. She tried to stop the bleeding,' João remembered. 'He was delirious and it took him [four] hours to die . . . it was an awful scene.'

Dornellas cut an exhausted figure as she came to terms with her failed bid to save the miner, who would be buried a few hours later in an improvised rainforest grave. 'It's a war here,' she told Dom.

A few hours later – after chain-smoking Ye'kwana elders authorised a trip to one of the region's largest and most notorious gold mines – Dom and João set off up the Uraricoera to see the damage such miners were inflicting firsthand.

Accompanying them was Edmilson Estevão, a Waikás-born Indigenous activist who had flown into the village with Dom and João (and his crying nephew) the previous day. A mild-mannered 33-year-old, whose cheeriness belied a troubled past, Edmilson was born in 1986 and grew up at the height of the last major gold rush. As a baby, Edmilson lost his father when his boat sank after he went out to salvage the roof of an abandoned *garimpeiro* shack he hoped to reuse in their village.

'I remember it a little, as if it were a dream,' Edmilson said of that period when I met him in Boa Vista in August 2023, nearly four years after Dom's visit. He said that when he was a boy, it was rare for Indigenous men to get involved in gold prospecting. 'We didn't speak Portuguese so well back then,' he said of the Ye'kwana, who represent about 2.5 per cent of the Yanomami

Regrow and Self-Protect: Indigenous Defenders

territory's population – some 760 people. Though this was no longer the case, Edmilson admitted to Dom during their trip. An increasing number of young Ye'kwana men were working as boatmen, ferrying miners to their isolated pits and barges and providing logistical support. Some even ran their own mines.

'He was shocked,' Edmilson remembered of Dom's reaction to that revelation.

'Can I quote you on that?' the British journalist asked cautiously, concerned that doing so might get his Brazilian guide into trouble with peers who were collaborating with the often armed illegal miners.

'Yes, you can,' replied Edmilson, who said he was determined to sound the alarm over the crisis facing his people.

Two hours upriver from Waikás's small port, the group's speedboat neared its destination, a sprawling mining hub called Tatuzão that takes its name from the Portuguese word for an armadillo. On the day Dom visited, Tatuzão – normally a bustling riverine community, boasting bars, brothels, a pharmacy and even a church – was unusually subdued. A few weeks earlier, the federal police had launched a rare clampdown, burning the clandestine township to the ground before withdrawing, leaving a handful of disgruntled residents behind.

The curly handwriting in Dom's notebook grew increasingly unruly as he raced to jot down details of the eclectic cast he encountered in the rubbish-strewn camp.

There were the sullen miners who had flocked to Tatuzão from depressed rural backwaters seeking to improve their families' lot. 'I know it's illegal,' shrugged one, who gave his name as Bernardo Gomes but admitted real names were never used in the mines.

There was the bar owner, Antonio Almeida, who downplayed the damage the miners were doing to nature: '[There's] no way you can kill it all.'

And there were the partially clothed sex workers, who Edmilson recalled being fascinated by Dom's piercing blue eyes. 'Your eyes are so beautiful,' Edmilson remembered one tipsy sex worker calling out after the journalist stepped off the boat.

'Dom just stood there writing it all down,' Edmilson added with a chuckle.

A few minutes upriver, Dom and João found the mines themselves. In one, a small group of *garimpeiros* had already started prospecting again despite the recent crackdown.

'Three men toil waist-deep in mud with a hose jetting water under an uprooted tree. Mud pours down a rough wooden sluice, while black smoke belches from a deafening diesel engine: a hand-operated industrial hell amid the wild tropical beauty,' Dom would later write in *The Guardian*, alongside one of João's apocalyptic photographs. In the image, the trio of filthy miners can be seen scrambling through a giant, clay-coloured trench resembling a bombsite where the rainforest once stood.

From the air, João's drone captured an even more disturbing scene: a vast ochre stain that stretched several kilometres towards the horizon and was filled with stagnant pools of water and fallen trees. 'It was a scene from a war – a battlefield,' João remembered. 'The sort of place I'd seen in films about . . . World War One. Those trenches. All those holes. All that mud.'

An even more pernicious impact was invisible to the eye: the miners' widespread use of mercury, which has poisoned rivers and fish populations on which Indigenous communities depend, here and across the Amazon.

Regrow and Self-Protect: Indigenous Defenders

For all the destruction Dom and João saw that November day, Edmilson told me that, in retrospect, it had only been the start. Over the course of the next three years, thousands more miners poured onto Yanomami lands, using increasingly sophisticated technology to plunder billions of dollars' worth of gold and cassiterite, which is used to make tin found in mobile phones and other electronic devices.

When I returned to the region with one of Brazil's best-known Indigenous activists, Sonia Guajajara, in December 2022, as many as 30,000 miners were estimated to have invaded. Mining gangs, we discovered, had even built an illegal 75-mile road through the rainforest and were using it to smuggle heavy machinery deep into the Yanomami territory's jungles to accelerate their operations.

'That's the road to chaos,' Greenpeace campaigner Danicley Aguiar told me as the three of us swooped over that clandestine highway in his group's surveillance plane – an aircraft familiar to journalists who cover the environmental crisis in the Amazon. 'And this is the chaos,' Aguiar added, pointing to a gaping hole in the rainforest where yellow excavators had gouged a mine out of another of the region's rivers, the Catrimani.

Even worse, the environmental catastrophe had contributed to a horrifying public health emergency. Activists said hundreds of Yanomami children had suffered preventable deaths since the start of Bolsonaro's administration, thanks to a lethal mix of government inaction, corruption and the explosion of illegal mining, which had scared off health workers and helped malaria to spread. Organised crime groups, such as São Paulo's First Capital Command (PCC), were said to be tightening their grip on the region's gold mines with the help of heavily armed gunmen.

'It was a government of blood,' Júnior Hekurari Yanomami, a Yanomami leader involved in fighting the health crisis, told me when I paid him a visit in Boa Vista in December 2022.

Fortunately for the Yanomami, by then, Bolsonaro's four-year administration was nearing its end. A few weeks earlier, Bolsonaro had lost the presidential election to the left-wing former president Luiz Inácio Lula da Silva, who had vowed to reverse the environmental calamity caused by his climate-denying predecessor. 'We will put a complete end to any kind of illegal mining,' Lula told me on the eve of his victory, promising to make the climate crisis 'an absolute priority'.

One month after we flew over the Yanomami territory together, Guajajara became Brazil's first ever Indigenous minister, and Lula's government launched what was billed as a major offensive to drive the miners off Yanomami land, similar to the campaign in the early 1990s. The new president travelled to Roraima to denounce what he called a premeditated genocide committed under Bolsonaro. A police investigation was launched. 'It's impossible to understand how a country like Brazil neglects our Indigenous citizens to such an extent,' Lula told journalists.

Yet the problems facing the Yanomami – and, indeed, the Amazon as a whole – did not start with Bolsonaro and would not end with his political demise.

When I returned to Roraima, in August 2023, Lula's eviction campaign appeared to be losing steam, thanks to a chronic lack of resources and a glaring lack of interest from the army and air force, which had failed to enforce a no-fly zone security experts believed was crucial to shutting down the mines. Miners who had been lying low during the early months of the crackdown were

Regrow and Self-Protect: Indigenous Defenders

returning to the Yanomami reserve to dig up their equipment and reopen their mines.

I went to see Edmilson at the headquarters of the Ye'kwana association, Wanasseduume, in Boa Vista to hear his take on the situation and get an update on the cacao project he had taken Dom to visit nearly four years earlier. Edmilson spoke optimistically about Lula's clampdown and how special forces from the rejuvenated environmental agency, Ibama, had destroyed several of the region's most polluting mines. 'The river's clean now,' he said of the Uraricoera, which aerial photos showed had changed colour from an earthy brown to a healthier shade of chestnut. 'Before, we couldn't fish at night without using a torch. You couldn't see the bottom of the river. Now you can.'

But Edmilson worried what might happen at the next election if Bolsonaro or someone similar was returned to power. 'We don't know who will come along after Lula,' he said. And he was also troubled that young Ye'kwana men continued to be seduced into joining the miners, despite the efforts of the cacao project to give them other options.

For a four-day round trip, transporting miners or sex workers, the going rate was about 5,000 reais (about US$1,000). An Indigenous ferryman might make five such trips in a month. 'That's a lot of money,' Edmilson admitted, although it paled in comparison to the tens of thousands of pounds plane and helicopter pilots could make for smuggling miners, equipment and stashes of gold in and out of the mines.

Edmilson feared the growing involvement of Indigenous youth in mining was gradually destroying Ye'kwana culture and traditions. Young people were adopting the ways of the miners, drinking throat-burning spirits such as sugar cane *cachaça* rather than the traditional fermented drink known as *caxiri*. Violence was on the rise.

Activists and village elders urged Indigenous youth to stay away from the miners, but were often ignored, Edmilson lamented. So what was the solution? 'The solution is for there to be no mining,' Edmilson replied bluntly. 'If there's no mining, then they won't be able to get involved. If there is illegal mining, they will end up getting involved.'

History shows that this is far easier said than done in a region as vast and inhospitable as the Yanomami's mountainous, forest-cloaked home. When the Brazilian government launched a massive clampdown to remove the *garimpeiros* in the early 1990s, then president Fernando Collor de Mello took the bold step of setting aside more than 1 per cent of his country's total territory for the Yanomami, to the outrage of conspiracy-embracing military chiefs convinced that foreign agitators were attempting to incite an Indigenous separatist movement there to commandeer the Amazon. Environmentalists and Indigenous campaigners celebrated Collor's move – but within a few years the miners were back, realising the potential profits far outweighed the risks of getting caught.

Now, under Lula, government promises to liberate Indigenous communities from illegal mining gangs appeared to be falling short once again, with the *garimpeiros* defying government attempts to dislodge them from Yanomami lands. In other areas severely affected by illegal mining, such as the Munduruku and Kayapó territories, it was largely business as usual.

That bleak reality was leading Indigenous communities across the Amazon to the same conclusion: that in the absence of effective

Regrow and Self-Protect: Indigenous Defenders

and sustained state action, they needed to start protecting themselves. In order to do so, they were setting up their own patrol and self-defence teams such as Evu, the group Bruno Pereira helped found, on which Dom had been reporting when they were murdered. Evu's initials represented the group's full Portuguese name – the Javari Valley Indigenous Association Patrol Team – but were also a reference to Bolivia's first Indigenous president, Evo Morales, who Bruno had admired. Other Indigenous communities, including the Uru-eu-wau-wau in Rondônia and the Guajajara in Maranhão, had also set up similar self-defence groups, the latter calling themselves the Guardians of the Forest. The Yanomami and Ye'kwana still lacked such a group of their own – although Edmilson told me he thought it would be a good idea and was being considered.

In Roraima's east, towards its mountainous border with Guyana, residents of Brazil's second most populous Indigenous territory, Raposa Serra do Sol, had created such a group, called GPVTI. In August 2023, I decided to pay them a visit with João to see if their work might provide one of the answers Dom had been looking for.

One afternoon, as a ferocious sun beat down on Boa Vista, we set off from our hotel opposite a seven-metre aluminium statue honouring Roraima's wildcat miners and took the route 174 highway north towards the border with Venezuela – the same road *garimpeiros* used to reach one of the secret Uraricoera river ports they used to steal onto Yanomami lands. But rather than taking a turning west towards that territory, we cut east into Raposa Serra do Sol, an Indigenous territory almost the size of Wales that is home to about 26,000 people from five different ethnicities, the Macuxi, Wapixana, Ingarikó, Taurepang and Patamona.

The landscape here was dramatically different from the Yanomami territory: instead of bewitching, mist-covered expanses of rainforest and dramatic granite peaks, parched rocky tracks wound through sweeps of savannah and rolling hills inhabited by giant storks known as *tuiuiús* and wild nocturnal cats called *maracajás*. But while the topography was different, the territories faced similar threats. Activists in Raposa Serra do Sol said they were also under siege from illegal diamond and gold miners, and believed many prospectors had relocated there since Lula's Yanomami crackdown began in February 2023.

A ferocious rainstorm prevented us reaching our destination on the first day of our journey, forcing us to hang our hammocks at a dank, scorpion-infested school on the territory's western flank. Dom had stayed there a few years earlier while covering an Indigenous summit, but had not mentioned his eight-legged hosts.

The next morning, we arrived in Tabatinga, a hilltop hamlet where GPVTI's activists had gathered to tell us their story of resistance. They greeted us at a roadblock at the village's entrance that blocked a dirt road leading to one of the largest illegal mines. The men carried bows and arrows and wooden truncheons. The air reeked of alcohol: the result of scores of bottles of cheap liquor they had confiscated from Indigenous smugglers the previous afternoon and smashed on the grass beside the post.

'Please take [our] message around the world so people understand that our struggle isn't easy – it isn't easy at all,' the village's 81-year-old chieftain implored us. 'And things are getting worse.'

For the next week, we toured Raposa Serra do Sol's eastern fringes with the Indigenous patrol team. We watched a Macuxi shaman dab malagueta pepper into the activists' eyes in order to protect and strengthen them as they set off on a mission down the

Regrow and Self-Protect: Indigenous Defenders

Maú River. We sped after GPVTI's 'warriors' as they chased down a Guyanese smuggler who was using a quad bike to take a cargo of Brazilian booze over the border to Indigenous communities in Guyana. We saw them use drones to gather intelligence on the illegal mines they said were polluting their rivers and devastating their culture.

'Defending our territory is part of life and it's part of our culture,' one of the youngest patrolmen, Marco Antônio Silva Batista, told me, describing the patrol group as part of a centuries-old history of Indigenous resistance.

At a roadblock on the border with Guyana, a 21-year-old activist called Jedeão Pereira Batista said he was determined to resist miners he called 'enemies of nature'. 'We consider them poison,' he said as he showed off his checkpoint.

For all the enthusiasm of GPVTI's young activists, I felt conflicted as we drove back to Boa Vista. Clearly, such groups had a crucial role to play in deterring the environmental criminals invading their territories. They could provide intelligence to the under-funded and under-staffed government agencies whose job it was to protect such Amazon communities from destruction. They could furnish journalists in faraway cities with information and images that might draw attention to crimes being committed in such remote and under-reported regions. But theirs was highly dangerous work, as the murders of Dom and Bruno had laid bare. On our last day with GPVTI, Marco admitted that just a few weeks earlier, machete-wielding miners had tried to abduct him as he tried to gather intelligence on a quarry near Tabatinga.

'We're not afraid anymore. We're used to the death threats. We don't care anymore. If something happens, that's our job – we're just trying to defend our territory,' the easy-going 20-year-old insisted.

Not long before that, the group's 4x4 had been shot at as it travelled to Boa Vista. Each night, as we crawled into our hammocks in Tabatinga, GPVTI warriors armed with truncheons and sticks took positions around our lodgings to guard against a possible attack. Their devotion was admirable and humbling, but surely there was no way such groups could be expected to face down such huge, well-financed threats on their own.

Before flying home from Boa Vista, I called in on a woman who knows far more about wildcat mining than most, to ask her if she saw any solution. Marina Cantão dos Santos spent decades working in mines all over the Amazon – including the legendary Serra Pelada excavation made famous by the photographs of Sebastião Salgado – before opening a fish restaurant on the banks of the Rio Branco, frequented by tourists and members of Roraima's mining elite.

Over a lunch of tambaqui and cassava, Santos regaled her visitors with death-dodging tales of her *garimpeira* days. The 67-year-old claimed that on one occasion, she had survived a plane crash with a daredevil bush pilot nicknamed 'Pé na Cova' ('One Foot in the Grave').

As we prepared to leave, I asked Santos if she thought illegal mining could ever be stopped. She shook her head: 'Only by throwing them all in jail or killing them.' It was a dispiriting thought and I suspected she was right.

Back in Rio, I pored through Dom's notes looking for a dose of badly needed optimism – and read again his reference to his desire to interview the veteran climatologist and Amazon thinker Carlos Nobre. I decided to talk to Nobre myself. When I told him of my pessimism, he chided me politely.

Regrow and Self-Protect: Indigenous Defenders

'When I was your age, I was super pessimistic,' he said, remembering his first trip to the Amazon in 1975 and how he had felt utterly despondent about the future of a region the dictatorship seemed bent on wrecking. That pessimism accompanied Nobre for much of his career. He went on to become one of the scientists who pioneered the distinctly depressing idea of an Amazon tipping point, after which its rainforests would suffer an irreversible breakdown, with terrible consequences for humanity.

These days, though, Nobre claimed he felt upbeat. 'There are numerous studies that show when a person hits 65, they become optimistic. And since I'm 72 now, that's that . . . I just said: "I don't know how many years more I'm going to live, so let's be optimistic."'

Nobre's hopefulness was built around an ambitious blueprint for the sustainable development of the Amazon that he and his colleagues called 'Amazonia 4.0'. The initiative's website told visitors it was dedicated to creating 'unprecedented alternatives to reduce Amazon deforestation' and, in doing so, tackle the climate emergency.

'New paths must be plotted and we are working on it . . . we are working to defuse a bomb,' it declared.

According to Nobre's vision, the future of the Amazon hinges on people's ability to create a thriving Amazonian 'bioeconomy', which would provide sustainable economic alternatives to the kind of short-term destruction Dom had witnessed during his trip to Waikás and Tatuzão. Nobre believed this paradigm shift could only be achieved by abandoning a centuries-old development model which ignored the immense biodiversity of the Americas and the traditional knowledge of its Indigenous inhabitants.

To illustrate his idea, the Brazilian scientist invoked the Italian explorer Amerigo Vespucci, who would later lend his name to the Americas region.

Vespucci arrived in what is now Rio de Janeiro in the early 1500s and was astonished by the things he witnessed after encountering members of the Tupinambá people in the region's Atlantic rainforests. The Italian navigator later waxed lyrical about seeing 'extensive and dense forests, which are almost impenetrable, and full of every kind of wild beast' as well as 'a spectacular variety of herbs, plants, seeds and fruits'.

'If they were our property, I do not doubt but that they would be useful to man,' Vespucci wrote.

Tragically, Nobre said, such ideas had not caught on. For centuries, European colonisers had preferred to impose their own agricultural techniques and traditions on the region, rejecting its natural riches and the knowledge of the people who already occupied the land. 'When the Portuguese, the Spanish, the French and the English arrived in the Americas, it was the place with the greatest biodiversity on Earth,' Nobre said. 'And they destroyed it.'

That development model persisted to this day and had to be changed if the Amazon was to be saved, argued Nobre, who saw initiatives such as the Yanomami cacao project as small pieces in this new bioeconomy.

Further south in another Amazon state, Rondônia, Nobre said he was seeking $1.2 million in funding to build a chocolate factory in an Indigenous territory inhabited by a people called the Paiter Suruí. '[It would be] an absolutely top-quality chocolate factory – and it would be the first modern factory in an Indigenous village since the Portuguese arrived here 500 years ago,' Nobre enthused.

Regrow and Self-Protect: Indigenous Defenders

Nobre thought another piece of the puzzle was a project called the 'Arc of Reforestation', a massive Amazonian reforestation scheme involving the recovery of 24 million hectares of destroyed or degraded forest – an area the size of the UK – between now and 2050. Nobre was convinced such a restoration scheme could help create millions of jobs for Indigenous and traditional communities and illegal miners alike. 'It's a challenge. Of course it's a challenge,' he said. But he considered it absolutely possible.

Four years after Dom's visit, the activists behind the cacao project in Waikás and the surrounding region also saw signs of hope, with 15 villages along the Uraricoera river and five further south along the Toototobi now involved.

In November 2023, Ronei de Jesus Silva, an Amazon-born agronomist who is helping spearhead the cacao initiative, managed to fly into Waikás for the first time in two years. In the years immediately after Dom's 2019 visit, the presence of illegal miners meant the security situation had been too bad for Silva and his colleagues to visit. 'It was chaos . . . that was the year that the floodgates opened,' Silva said, remembering flying over Waikás on his way to other villages and catching a glimpse of the turmoil below. 'You'd see all the [mining] engines. Loads of boats. Mining going on really close to the community.' But Lula's albeit incomplete crackdown on the miners had restored a measure of normality and safety. Silva said he was encouraged by what he saw in Waikás – and believed Dom would have been too, had he still been alive.

In the four years since Dom passed through, villagers cultivated three groves with about 3,000 cacao trees. They had yet to make any money from the initiative but hoped to start commercialising the fruit after the 2024 harvest.

'The community is picking itself up . . . and this is really good to see,' Silva said. 'Things are better now. Many wounds were inflicted – but nature will heal.'

Another person I wanted to hear from was Davi Kopenawa, who had been relentlessly campaigning for his people's survival for as long as I had been alive. In many ways, it seemed to me, Kopenawa's life story provided an answer to Dom's question of how to save the Amazon. Without his tireless struggle and globe-trotting activism, the Yanomami territory might never have even been created in the first place. Perhaps his people might already have been condemned to the history books, as had happened with so many other Indigenous groups in the five centuries since Portuguese explorers claimed to have 'discovered' Brazil.

When we spoke in late 2023, Kopenawa, who, like Nobre, was now over 65, also exuded optimism, despite his growing frustration that *garimpeiros* were returning to the Yanomami territory as Lula's eviction campaign faltered. A few weeks later, the 67-year-old Indigenous icon would fly to Rio to take part in its annual carnival procession with a samba school that had devoted its parade to championing the Yanomami cause and Kopenawa's decades of resistance.

The samba's refrain was a Yanomami cry of defiance – '*Ya temi Xoa!*' which means 'I am still alive!'

At the end of our conversation, Kopenawa invited me to visit his territory to witness the upsurge in destruction – just as Dom had done four years before. 'I need you to see this up close,' Kopenawa said, vowing to show us 'the big machines that have made the people of the forest sick'. We were keen, though sadly our plan was thwarted when the Brazilian military refused us permission

Regrow and Self-Protect: Indigenous Defenders

to perform a series of flyovers during which we hoped to document the situation.

Kopenawa said he believed journalism was also part of the solution to the problems blighting his people's territory and the Amazon as a whole. 'You journalists use a different weapon, the weapon of announcing the news . . . to those who are unaware of our suffering,' he said. 'So I hope that you journalists, who live so far from my community, will carry on. I want you to continue pressuring the Brazilian state . . . so that it respects us, for all of our sakes,' he explained.

It was an exhortation that resonated deeply with me, particularly after Dom was murdered while doing precisely that.

'What I do know is that we must continue fighting together. We are together and we must continue fighting together,' Kopenawa said of the activists and journalists engaged in the Amazon. 'These invaders who threaten our lives will not stop. They will carry on abusing [the Amazon]. And you and we, the Yanomami people, we will carry on speaking out.'

Costa Rica

CHAPTER SEVEN

Putting a Price on the Future: Tourism and Environmental Service Payments

Stuart Grudgings and Dom Phillips
Costa Rica

Stuart Grudgings is a former Reuters bureau chief who covered politics, economics and the environment for two decades as a reporter and editor, including assignments in Japan, the Philippines, Brazil, Malaysia and Washington. He wrote extensively about threats to the Amazon rainforest and potential solutions during his time as a Rio de Janeiro-based correspondent, from 2008 to 2011.

'We realised we had a huge market failure.'

Víctor Merella padded up the steep jungle trail with the full sensory presence of a seasoned hunter. 'Look: an ocelot was here,' he whispered, pointing to some faint, feline paw prints in the mud that I was about to tramp through.

This was one of many signs we saw, during an hour-long trek through dense, muddy rain forest in Costa Rica's Osa Peninsula, that wildlife was thriving in a jungle that was cattle pasture only 20 years ago.

Merella, a youthful 40-year-old with a long ponytail, had already pointed out fresh tracks indicating the presence of several other mammals, including agoutis (a large rodent native to Central and South America), raccoon-like coatis and tapirs. Somewhere nearby was a pack of white-lipped peccaries, a type of jungle boar known locally as *chanchos del monte* (mountain pigs). Their meat is prized throughout the region and they are an increasingly threatened species.

The forest around us was a hunter's paradise. But Merella hasn't killed a forest animal for food in a decade. He works as a wildlife guide and monitor, taking tourists on day and night jungle walks and helping to run a Belgian-funded project studying and protecting the venomous bushmaster snake. Merella and his community, Rancho Quemado, have undergone a remarkable transformation from forest destroyers to jungle protectors in the last few decades, one that mirrors a change in Costa Rica as a whole.

'We used to kill everything that moved,' said Merella. He grew up here hunting animals and felling trees to make way for pasture in the 1980s and 90s.

Even the community's name, which means 'burnt ranch', is intimately linked to forest exploitation. The story goes that early settlers accidently burned down their ranch when fat from a stove loaded with hunted meat sparked a fire.

Costa Rica was an essential stop for Dom as he searched far outside Brazil for lessons that could help the Amazon. He traversed the country in early 2021, speaking to officials, farmers and environmentalists. And he also trekked through the jungle with Merella to try to understand the social, economic and political forces that can persuade a community to transform its relationship with the forest. Speaking to Costa Ricans in a mix of broken Spanish and fluent Portuguese, Dom sometimes expressed amazement tinged with exasperation at just how different attitudes to the forest and the broader environment were to those he was used to.

'In Brazil, the people just see the forest as a place to hunt and raise cattle – they don't have this way of thinking,' he told Merella after the former hunter explained how the Rancho Quemado community now teaches conservation to kids to ensure a sustainable future.

Costa Rica, a Central American nation of 5 million people, is the world's deforestation turnaround specialist – a title it's earned while building a successful, stable economy fuelled by tourism. That was barely imaginable just four decades ago, when Merella was born. Much like Brazil has in recent years, Costa Rica was destroying its rainforest at one of the fastest rates anywhere. Inspired by the land reform ideals that spread through Latin America in the second half of the last century, it had adopted a formal policy in 1961 of expropriating and redistributing land that was deemed unproductive. The goals were to push the agricultural frontier, expand production and grow export markets. The government used agricultural subsidies and price guarantees to boost agricultural output, while banks incentivised farmers to clear more land in exchange for credit. Little thought was given to the environmental impact of any of those policies.

The 700-square-mile Osa Peninsula, which juts into the Pacific along Costa Rica's southern coast, was in many ways ground zero. The region has been called the most biologically rich place on earth, home to 2.5 per cent of the world's biodiversity, from the jaguars that roam its Corcovado National Park to close to 700 species of trees to the humpback whales that nurture their young in its protected bay. Cattle pastures on its outskirts steadily give way to dense, uninterrupted rainforest nestled against largely deserted stretches of beach.

But from the 1950s through to the 1990s, the forest was disappearing rapidly under pressures that are familiar in the Amazon today. A US-owned company, Osa Forest Products, bought up 47,500 hectares in 1957 and set to work clearing forest and dredging rivers for gold. Thousands of pioneering families arrived to carve homesteads out of the ancient rainforest and pan for gold in its rivers.

Marjorie Gamboa Vargas, a 63-year-old farmer who was a teenager when her family migrated south to the region, remembers when the country's then president visited and distributed formal land titles to settlers. Her 27-hectare property on the outskirts of the national park – about an hour's drive from Rancho Quemado – has less than a fifth of its original forest cover intact. The rest is cattle pasture interspersed with patches of trees on gently rolling hills.

'The land is yours! Costa Rica needs rice, beans, and corn,' she remembers President Rodrigo Carazo, who was president from 1978 to 1982, telling settlers. 'He told us to get to work. We thought that the further you could go into the forest, the better.'

By the late 1980s, the cost of these policies had become apparent throughout Costa Rica. As more and more land was converted to cattle ranches and plantations growing crops for export, including coffee and bananas, indigenous species of plants and animals were vanishing, and the country's once rich soil was fast being degraded. Costa Rica lost at least 30,000 hectares of forest per year in the 1970s and 80s. By 1986, this had reduced the amount of forest cover to about 25 per cent, down from over 70 per cent in 1950.

But in the late 1980s, there was a shift. Public attitudes toward conservation began to change and so did the political discourse.

The shift was an outgrowth of a radical move Costa Rica had made decades before, in 1948, when it abolished its army and began to pour money into education – over 6.5 per cent of GDP, among the highest commitment in any OECD nation. The country had a literacy rate of 98 per cent in 2021, four points above the Latin America and Caribbean average. Curriculum about biodiversity and conservation was taught in schools.

And Costa Rica is a country with a long history of democratic stability, adherence to the rule of law and working towards public consensus on challenging issues. There was space for a thoughtful public conversation about the use of natural resources, which was influenced by global concern for rainforests. Political leaders began studying a radical idea – putting an economic value on nature. They looked first to the national park system.

In 1970, Costa Rica's government had begun to create an extensive system of national parks, inspired by that in the US. The 164-square-mile (42,500-hectare) Corcovado Park (including 13 square miles (3,370 hectares) of protected marine reserve) was founded in 1975. Tourists were drawn to the parks' waterfalls, beaches and abundant wildlife. And the parks grew: by 1990, about 17 per cent of the country's territory was protected, among the highest proportions in the world. And the contours of an alternative model for using forests was emerging.

Carlos Manuel Rodríguez, who served as environment minister three times, told Dom that the increasingly apparent costs of the old policies – including declining biodiversity and soil quality – made policymakers realise that market forces weren't serving Costa Rica well. 'We realised we had a huge market failure,' said Rodríguez, who was the country's director of national parks in the early 1990s. 'The market was unable to reflect the benefit of forested areas to humans, to Costa Rica, to the economy.'

The country was operating without a map – no other tropical nation had successfully reversed deforestation. The country's Tropical Science Center conducted a pioneering valuation study, putting a price on over two dozen environmental services, including water production, climate benefits, biodiversity and soil conservation. FUNDECOR, an NGO established to promote sustainable development in Costa Rica's central volcanic region, began a programme making small payments to farmers who chose to conserve forest instead of cutting it. Its success paved the way for the adoption of payments for environmental services on a national scale – the foundation stone of Costa Rica's remarkable reforestation story.

The shift in public opinion became apparent with the 1994 election of President José María Figueres, who made conservation central to his four-year term. Lawmakers then passed the landmark 1996 Forest Law, banning the conversion of remaining forest land to other uses and recognising four key environmental services provided by trees: carbon storage, water, biodiversity protection and scenic beauty. Under the law's Payment for Environmental Services programme, known as PSA for its Spanish acronym, landowners could apply to an agency for yearly payments in return for their provision of those services.

In 2023, that amounted to the equivalent of around US$70 per hectare per year. It doesn't sound like much. But GDP per capita in Costa Rica is US$16,600 a year and significantly less than that in rural areas. So those payments have been enough to deter deforestation. To fund the payments, the government adopted a 'you pollute, you pay' principle, imposing a national tax of 3.5 per cent on fossil fuels.

'We know that the worst enemy of conservation is poverty,' Mario Piedra, the head of FUNDECOR, told Dom. 'The poorer you are, the more prone you are to liquidate your assets. It's just human nature.'

By 2021, the programme had invested over US$500 million in contracts covering more than 1.3 million hectares – around a quarter of the country. And the transformation it sparked is vividly apparent when you compare satellite images of the country taken in 1997 with images from 2015. Large tracts of parched yellow and brown landscape in the earlier images give way to dark, lush greenery reclaiming much of the country. Total forest cover has rebounded to around 60 per cent.

The single greatest dividend from the incentive payments and Costa Rica's extensive national park system has been tourism. The country has become a byword for pristine, unspoiled nature, pulling in over 2.75 million tourists in 2023 – accounting for 8 per cent of GDP.

'Ecotourism became a major driver' of reforestation, said Rodríguez, who played a leading role in implementing the payments plan and is now head of the US-based Global Environment Facility, a multilateral environmental fund. 'It's the main source of jobs in rural areas.'

The PSA also became an important tool for social inclusion. Female landowners get preferential access to the payment system, as do the country's Indigenous people, of whom there are about 100,000. Indigenous peoples such as the Bribri, Cabécar and Boruca, most of whom live in the central and southern highlands, have long been land rich and cash poor.

In the southern highland territory of Salitre, Bribri association leader Guillermo Elizondo told Dom that 5,000–6,000 hectares – half the Indigenous people's land – was covered by PSA contracts. That works out to total payments in the region of US$385,000 per year. For the Bribri, Elizondo said, the true mandate to protect nature comes from their god, Sibu, who is 'in everything – the rain, the sunrise, the stars'.

'PSA is not their reason for conservation,' said Elizondo, then aged 50. 'It has only been here for 15 to 20 years. We've been conserving the land for thousands of years.'

But PSA places a hard economic value on the Bribri's conservation efforts for the first time. The payments fund the community association's entire budget, including education, housing, water and electricity.

The PSA is not, as Dom learned, a perfect system. As it has grown, its flaws have become apparent and some Costa Ricans have become frustrated by it. Andrea Herrera used to work as a forest 'regent', or engineer – the officials who confirm details of PSA contracts with landowners and monitor that the terms are being met. Lots of bureaucratic red tape and high administrative costs make it hard for lower-income landowners with smaller plots to access the programme, she said. It's common for families to spend hundreds of dollars applying for PSA funds, only to be rejected or face delays in approval that can stretch to a year, she said.

'The cost of applying to PSA is high,' said Herrera, who is now the executive director of an NGO that supports community conservation efforts and helps landowners get access to the payments scheme. 'It's not worth it for people with one or two hectares.'

Marjorie Gamboa Vargas, the farmer in Osa, agrees. The four to five hectares in untouched forest she has on her 27 hectares of land would only fetch her US$280–350 a year in payments, an amount she said isn't worth the bureaucratic headaches and application costs.

Still, it's telling that when Costa Ricans debate the PSA programme, they mostly talk about how it could be improved, not about scrapping it. The discussion is about how access could be expanded and how it could be paired with other conservation schemes.

The payments programme helped drive reforestation, but Costa Rica took another key step that has helped maintain the thriving biodiversity in national parks and dozens of other protected habitats, from mangroves to volcanic highlands to forest reserves. Government partnered with NGOs and local communities to establish a network of 44 biological corridors that run from north to south and from the Atlantic to Pacific coasts. These forest pathways allow 'keystone' mammals like peccaries, tapirs, monkeys and jaguars to traverse different biospheres, spreading the seeds of crucial tree and plant species as they go.

One of the most ambitious of these corridors is being built in Osa with the help of farmers and communities such as Rancho Quemado. Yolanda Rodríguez, the 47-year-old head director of Rancho Quemado's monitoring committee, described how during her childhood, the most common sighting of a white-lipped peccary was when they were slung lifeless over the back of a hunter's horse.

The shy, boar-like mammals move in packs of up to several hundred, performing a vital ecosystem engineering role as they snaffle up seeds and spread them along their migration routes. But intense hunting had driven them deep into the forest and they had virtually stopped appearing near Rancho Quemado until the community began a major effort to protect and monitor them in 2018. Now, more than 300 *chanchos* have a permanent presence in the area, with many more making regular migrations to and from the Corcovado Park, Rodriguez told me in the kitchen of her simple house.

Tears welled in her eyes after I told her of Dom's murder in the Amazon. She pointed out a small tree in her garden; Dom had interviewed her in its shade two years earlier.

The community's white-lipped peccary protection programme and its other conservation efforts have received funds from a US$26 million 'debt for nature' swap – an agreement to write down external debt in return for an internal commitment to invest in conservation – signed between Costa Rica and the US in 2007. They're also sustained by scientific partnerships with NGOs and universities, as well as by a trickle of foreign ecotourists and bird-watchers. Asked what was the key to Rancho Quemado's success, though, Rodriguez identified two non-financial factors as the most important – community organisation and love.

'We have to love what we're doing,' Rodriguez said. 'If this environment where my community and family are living isn't healthy, they won't be healthy either. Love is the key.'

For working farmers in the region, however, conservation has to have a more direct economic payoff. Cira Sánchez Sibaja, 58, inherited her 370-hectare farm in the Osa peninsula when her father died two decades ago. But she never spent much time there until a cancer diagnosis prompted her to leave the capital city San José in search of a healthier life. When she began the transition in 2008, it was daunting: she had no farming experience, but now she had to manage an endless horizon of cattle pasture. The farm had no electricity or internet access.

'I thought I was crazy. What could I do?' she said.

Putting a Price on the Future

She ended up thriving by taking a different approach to the land than her father's generation. She talked to local conservationists who convinced her of the potential for creating more sustainable, productive cattle-raising practices while also attracting ecotourists.

In contrast to the wooden posts and barbed wire that contain cattle in most farms here, many of her fields are surrounded by trees – so-called 'living fences' that keep cows in while providing shade, protecting water sources, giving birds a nesting ground and mammals a corridor to pass through.

With the help of Osa Conservation, an environmental non-profit group, she plans to plant 40,000 trees in the next few years on four hectares set aside for reforestation. Income from tourists gives her another incentive to conserve. Today, she can accommodate up to 20 visitors in simple cabins, offering them kayak tours of the Golfo Dulce (Sweet Gulf) bay and horseriding on the beach.

'I still have just as many cows but I added value,' said Sebaja, a stocky, energetic woman who was wearing jeans and a cowboy hat. 'I didn't need to stop any activity.'

Conservation in a poor, rural area like Osa can only work if it is focused primarily on raising people's production and incomes, she said. 'Otherwise, what will people eat? Leaves?'

In his interviews, Dom was consistently struck by how Costa Rican farmers like Sebaja – while often sceptical of government initiatives and far from being tree-hugging environmentalists – were able to see the long-term value of conservation over shorter-term economic gains. 'Farmers talking about the importance of conservation rather than just paying lip service is so different to Brazil,' he said during one interview. 'In Brazil, it's just a legal requisite they have to follow.'

As Costa Rica runs up against the limits of how much forest it can protect outright, environmentalists are having to find ways to promote conservation and biodiversity within the framework of local economies. In Osa, that means securing the trust of landowners and providing educational and financial support that helps them see this long-term productive value in conservation.

Winning the hearts and minds of farmers like Sebaja and Vargas is central to Osa Conservation's 'Ridge to Reef' vision. The goal is to restore degraded forest and thereby increase species connectivity from Corcovado to the country's central Talamanca mountain range, about 80km to the east. In three years, it's recruited 346 farmers into its restoration network.

'We don't go to farmers and talk about the environment,' said Jose David Rojas, the group's lowland restoration coordinator, who trained as a forest regent. 'We talk about productivity first and we listen to them.' Rojas' team works closely with agriculture ministry officials to present a plan to farmers that offers ways to improve productivity, boost resilience – and promote conservation. That might be in the form of planting trees to protect water sources and avoid erosion, creating living fences for cattle, or using techniques such as rotating cattle grazing to reduce land degradation. Or it could be helping them diversify their production by planting vanilla and cacao trees that will produce income while still promoting biodiversity and connectivity.

The group has planted more than half a million trees in recent years and now has five nurseries where it grows more than 100,000 native trees a year for use in restoring degraded areas. This also serves the purpose of dispersing rare and endangered tree species over a broader area.

Costa Rica's government makes it relatively easy for environmental groups like Osa Conservation to do their work and often

Putting a Price on the Future

actively partners with them – not always the case in developing countries. Carolina Soto-Navarro, the group's then director of conservation, previously worked in Southeast Asia, where, she said, there was very little government support for conservation and international NGOs. Brazil has a long history of hostility towards such groups working in the Amazon, suspecting them of being a front for foreign interests seeking to take control of its resources. 'All of our operations [in Costa Rica] are possible because we have very, very good relationships at the government level,' Soto-Navarro said.

Osa Conservation is piloting a programme that allows farmers who meet biodiversity targets – such as soil and water quality, and bird population numbers – to get relief on loans worth between US$5,000 and US$15,000 over a three-year period. Andrew Whitworth, the executive director of Osa Conservation, says biodiversity projects like this help address a key weakness in Costa Rica's environmental record – its continued heavy reliance on monoculture export crops such as pineapples and palm oil. These industrial-scale monocrops may count as forest cover for the PSA – stands of teak trees, for example. But they reduce biodiversity and their production relies on heavy use of chemical pesticides – Costa Rica has one of the highest intensities of pesticide use in the world.

'The PSA was great: it increased forest cover but a lot of that forest cover isn't necessarily good habitat for wildlife,' Whitworth said. Promoting more biodiversity isn't only good for wildlife, he added – it helps farmers become more resilient to the rollercoaster rises and falls in global commodity prices. Indeed, there's a growing body of evidence that blending trees with farming, or agroforestry, is a win–win for agricultural production and the environment. A 2020 study of agroforestry in sub-Saharan Africa, for example, found significant gains in crop yields at farms that used practices such as crop rotation. Trees' cooling shade and the protection that their root systems provide against soil erosion help build resilience against climate change. An 18-year-long study in Costa Rica published in 2023 found that bird species were able to thrive in small farms with natural landscape features, in sharp contrast to how they fare in monocrop plantations.

The PSA programme and other efforts to win support for conservation in communities around national parks have been integral to Costa Rica's success. Tree cover and biodiversity have expanded around its parks over time. That's a sharp contrast to the situation in Costa Rica's Central American neighbours such as Nicaragua and Honduras, and to much of the Amazon, where protected areas have been steadily encroached upon by human activity – from agriculture to illegal hunting and fishing to tree clearing.

A 2012 study published in the journal *Nature* found that environmental changes immediately outside tropical forest reserves 'seemed nearly as important as those inside in determining their ecological fate'. Changes inside reserves strongly mirrored those occurring around them, it found. 'It doesn't matter if you make protected areas if you don't invest in the landscape and people and the economic opportunities immediately around them,' said Whitworth.

With the PSA scheme often offering little more than a marginal incentive, the key economic opportunity supporting conservation in Costa Rica is its thriving tourism sector. When Dom visited, the country was just beginning to emerge from the painful economic impact of a pandemic that forced it to close its borders to international visitors for more than four months. Tourist arrivals that year plunged to 1 million from over 3 million in 2019, and only reached 1.3 million in 2021.

Putting a Price on the Future

'It was like someone came and turned the lights off,' Dionisio Paniagua, a 48-year-old who had been a jungle tour guide in Osa for 20 years, told Dom. He said the pandemic had been a loud wake-up call on the need to double down on conservation. 'The whole area was really scared,' he said. 'We realised that tourism is the heart of the economy.'

Even if people weren't directly employed in tourism, they didn't escape the impact as spending across the local economy slumped, he said. 'If we lose biodiversity, we won't be able to attract tourism. Other places have forests and better beaches.'

Tourism offers a way for landowners in areas such as Osa to supplement and diversify their incomes, and to employ more people. Sebaja said her accommodation and activities for guests provide work for six local people, compared to the two people she employs on her farm.

Dom spoke to Pedro Garcia, a 58-year-old landowner in the northern district of Sarapiqui, whose family moved there as part of a wave of squatters seeking land in 1984. In the year before the pandemic, Garcia said, about 80 per cent of his income came from tourists who toured his 3.5-hectare farm to learn how to plant trees and protect the forest within a working farm.

Aided by an incentive payment of about US$2 per tree from the PSA, and advice from FUNDECOR, Garcia had planted over 1,000 trees in recent years, including cacao and the endangered mountain, or yellow, almond species. 'I have a productive forest,' Garcia told Dom. 'I want to show that you can produce and protect at the same time.'

Costa Rican farmers are not all committed conservationists: many are as motivated by economic returns as the ones Dom had met in Brazil. They voice scepticism at government environmental policies and are adamant that protection for protection's sake isn't a viable path.

'Shouting and punishing doesn't work. We need more incentives,' Vargas told me. Vargas is working on a plan with Osa Conservation to plant over 1,000 trees on her farm and develop an agritourism project, but says she wouldn't hesitate to cut down a tree 'if I need it to survive'.

The difference is that Costa Rican landowners are able to take a longer-term view, knowing they have support – whether through the PSA, tourism dollars or help from groups like Osa Conservation – to put the time and money into establishing conservation-minded practices. Added to this, a consensus-driven, technocratic approach has helped Costa Rica achieve a policy coherence and consistency on the environment that many countries – notably Brazil – have lacked. 'There's a very strong middle ground that helps us build on what the previous government did,' said Rodríguez, who himself was brought in to help create the PSA programme despite being in the opposition party.

Costa Rica has broken down many of the silos that often stifle coordination on environmental policy. Its conservation, energy and mining policies fall under a single ministry. SINAC, the agency responsible for national park management, combines three previous organisations that had managed laws relating to national parks, wildlife and forestry.

'We need to recognise that we have a systemic challenge, so the response should be systemic,' said Rodríguez. That means having institutions that work at the 'landscape or seascape' level as opposed to having conflicting agendas, he explained.

Yet despite all that Costa Rica has achieved in restoring and protecting forests, there is a growing recognition that what worked here in the past won't be enough for the future. Several big threats to the model have emerged. Ironically, one of those is the national goal to become a net-zero, decarbonised economy by 2050. The country's electricity supply already comes almost entirely from renewable sources, mostly hydropower. But any visitor to Costa Rica can see where the decarbonisation needs to happen – on its roads. Transportation is by far the biggest source of greenhouse gas emissions; San José snarls with traffic jams and the country's roads are full of older models of trucks and cars.

But phasing out conventional vehicles, in favour of more public and electrified transport, works directly against Costa Rica's key environmental strategy. The PSA programme gets its funding from a tax on fossil fuels. If less of those are sold, there will be less in the fund to pay incentives to landowners. The Covid pandemic provided a taste of this impact. The sharply lower demand for fuel during lockdowns drained PSA finances, leaving the agency that administers the fund scrambling to meet demand from landowners. In 2023, landowners applied for contracts to cover 174,793 hectares with the programme but the budget would only extend to 69,283 hectares.

'In a decarbonised economy, we don't have the luxury to seal off half the country,' FUNDECOR director Mario Piedra told Dom during his visit.

Piedra said it was crucial for Costa Rica to create alternatives to tax-funded programmes that would give landowners stronger and more sustainable incentives to conserve forest. PSA payments have barely changed in 25 years, while the potential profits from cattle and pineapples – of which Costa Rica is the world's biggest exporter – have soared, he noted. Between 2015 and 2019, Costa Rica lost 1,200 hectares of tree coverage due to pineapple farming activity, according to a recent study, adding to around 5,600 hectares lost in the 15 years from 2000.

'When we say we want to give value to the forest, it's to give financial value – not only feelings and emotions and principles,' Piedra said. 'We shouldn't forget we are a middle-income country and that return on assets is important.' Allowing land owners to harvest trees sustainably is one way of doing this, he said. FUNDECOR has expanded a programme with the Germany-based Forest Stewardship Council that allows owners to cut down a certain number of trees based on a detailed sustainable forestry plan. The results have shown that the forest can become richer and more productive by harvesting some larger trees with big canopies and allowing others to take their place, he said.

But Piedra's eyes are on a bigger prize – the coffers of the world's largest companies. FUNDECOR has partnered with a technology firm to create the BIOTA project, which offers blockchain-based investments in tracts of forest to corporations and governments seeking to offset their environmental impact and meet corporate sustainability targets through biodiversity credits.

The concept of biodiversity credits got a significant boost at the COP15 United Nations Biodiversity Summit in late 2022, where countries made a '30 by 30' pledge to protect 30 per cent of land and water by 2030. The goal of the BIOTA project, Piedra said, is to turn 'an environmental service into a product that people can invest in'. Done well, it provides a far truer measure of a forest's contribution to ecosystem integrity than the simple per-tonne metric on which traditional carbon credits are based, he said.

Each Costa Rican 'product' on the BIOTA site provides a detailed breakdown of the land's biodiversity assets and value. An entry for a 354-hectare tract called Ara Ambigua in northern

Costa Rica, for example, cites its average annual rainfall, its role as a bulwark against local urbanisation and agricultural expansion, and how it provides crucial protection for threatened species like the great green macaw. Piedra's aim is to win investment in 100,000 hectares over the next few years by 'going to the people who have money' – aiming to have big multinational corporations invest in the largest tracts of land, and local companies in smaller parcels. He estimates that annual payments to landowners will be around US$1,300 per hectare after administrative expenses are accounted for, vastly more than the PSA's US$70. (Only land that isn't covered by other conservation incentive programmes will be eligible for BIOTA funds.)

Another track through which Costa Rica receives foreign funding for preservation is international carbon credits. Yet the returns are relatively small. A US$60 million agreement signed by Costa Rica and the World Bank in 2020, for example, only translates to about US$7.50 in annual payments per hectare. Piedra dismisses carbon credits as 'peanuts'. 'The forest should be a trust fund for these families,' Piedra said. 'Almost like having an oil well in Louisiana.'

Yet just as Costa Rica's reforestation efforts are crying out for fresh support, the political consensus that has sustained its environmental success appears more fragile than ever. A long-running fiscal crisis that was worsened by the pandemic, together with rising levels of unemployment, poverty and violent crime, has pushed the environment down the government's agenda.

The 2022 election of President Rodrigo Chaves, who channelled anger against the political class, marked a turn towards right-wing populism familiar to Brazil and the United States. Chaves has voiced support for exploiting natural gas deposits, a previously taboo subject, saying it could help transform Costa Rica into a 'new Singapore'. He kicked off his presidency by cancelling an intercity light rail project and Congress then pulled Costa Rica out of the 2018 Escazu Agreement that requires Latin American governments to promote the public's right to participate in environmental decision-making.

At the same time, several regions of Costa Rica are struggling to control the environmental and social impact of their popularity among foreign visitors. The very natural beauty that draws tourists and long-term foreign residents to the country is being threatened by the often poorly controlled development that springs up to serve them.

The small Pacific Coast town of Uvita where I have lived since 2018 has experienced a boom driven by foreign spending. The lush, forested hills in the area are increasingly pockmarked by the flattened red earth of lots carved out of the jungle to make way for luxury ocean-view homes that sell for US$1 million and up – far beyond the reach of most locals. The rise in property values and rental prices to North American levels has made it tough for local Costa Ricans to live near where they need to be for work. The foreign-driven gentrification has fuelled robberies and burglaries as the income gap has grown.

It's also having a worrying impact on local ecosystems, said Pablo Piedra, a 45-year-old former diving instructor who's lived in the area for two decades. The loose earth created by housing development is swept into the rivers by rain, creating ever bigger clouds of sediment that suffocate coral and kill other marine life, he said. The lack of a centralised sewage system means every home, restaurant and hotel requires a septic tank, which leaches human waste into waterways, increasing health problems and harmful algae blooms in the ocean.

Putting a Price on the Future

Piedra, who founded and heads the Uvita-based Costa Rica Coral Restoration group, said that some of his favourite dive spots – including Uvita's iconic national park beach shaped like a whale's tail – have been ruined by the changes. 'Not so long ago, I was able to go diving at the Whale Tail when I was training people for open water courses,' Piedra told me. 'Now it's nothing – there are still some fish but no coral. We lost all the coral that was there.'

The problem, say Piedra and other local environmentalists, is the lack of any overarching plan to monitor the impact of development and enforce limits. It's become common practice for developers to skirt environmental requirements by building without a permit, knowing they face a small risk of fines. Setting fires on land to destroy forest and convince authorities there was no unnatural change in land use is another favoured tactic. In the nearby community of Ojochal, a huge week-long fire in 2019 destroyed dozens of hectares of forest on land where a large property developer plans to construct hundreds of houses.

As Costa Rica reaches a pivotal point, it's worth remembering how it all started early last century – with widespread, massive destruction of its forests in pursuit of economic gain. Reforestation isn't easy or cheap. An economic analysis conducted by Osa Conservation found that the cost of fully restoring the ecosystem from the peninsula to the highlands would come to US$100 million over 10 years.

'That's the challenge,' said Andrew Whitworth, the group's director. 'How do you avoid destructive extraction and get ahead of the game? Don't do what Costa Rica did. It's a lot cheaper to protect and prevent something from being broken than to try and repair it afterwards.'

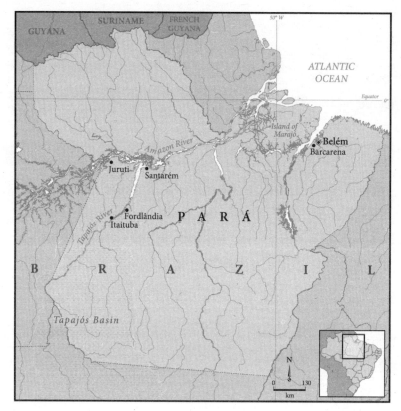

Tapajós Basin

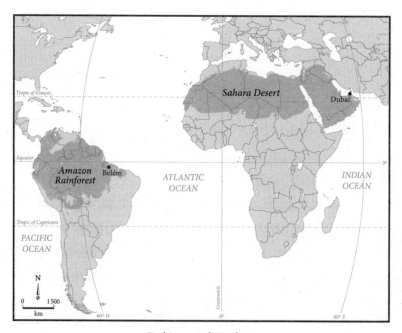

Belém and Dubai

CHAPTER EIGHT

SHAKING THE GLOBAL MONEY TREE: INTERNATIONAL FINANCE

Andrew Fishman and Dom Phillips
Tapajos Basin, Pará State, Brazil, and Dubai,
United Arab Emirates

Andrew Fishman is the president and co-founder of The Intercept Brasil, *an award-winning, reader-funded investigative newsroom. He reported extensively on the Edward Snowden archive and the Vaza Jato leaks, which exposed fraud and illegalities in Brazil's influential Operation Car Wash task force. He previously worked at The Intercept (US) and National Public Radio, and has been published in* Al Jazeera, Jacobin, Agência Pública *and* Todo Notícias, *among other outlets.*

'Man's conquest of Nature turns out, in the moment of its consummation, to be Nature's conquest of Man' —
C.S. Lewis, The Abolition of Man

The wily and irascible automobile magnate Henry Ford was looking for a cheaper source of latex to reduce the cost of Ford motorcars that were rapidly transforming the landscape of the United States. Almost a century ago, to great fanfare, he set out to construct Fordlândia, his ambitious vision for a million-hectare capitalist utopia on the banks of the Tapajós river, a major tributary that meets the Amazon in the Brazilian state of Pará. Ford's men broke their backs clearing forest, while dodging poisonous snakes, and meticulously planted row after row of rubber trees to turn the wild forest into a green factory of sorts that would print money for a foreign industrial empire.

The plan never really made much economic or ecological sense, as some internal critics pointed out at the time, according to historian Greg Grandin's seminal book on the town, *Fordlandia: The Rise and Fall of Henry Ford's Forgotten Jungle City*. But Ford, confident in the superiority of his American gumption, values and technology, felt he was doing God's work by conquering the forest and bringing 'civilisation' and economic progress to what he saw as the backwards peoples of the jungle.

Except the Amazon did not lack Ford's brand of rubber plantation due to ignorance. Just the opposite. As the locals knew, pests and blights quickly spread between native rubber trees planted close together. Despite 18 years of intense effort from one of the world's pre-eminent enterprises, barely any rubber was ever produced. The industrialist died having never set foot in the Amazon and Ford's grandson wrote off a loss worth US$350 million today. The forest won.

Ninety-five years after the town's founding, a Fordlândia resident sitting on the riverbank could watch the inaugural voyage of a ferryboat slowly pushing 35 long and slender white covered barges lashed together, seven by five, upstream.

They were on their way to a port in nearby Itaituba operated by American soy trader Cargill, where they will be loaded with 70,000 tons of grains for buyers in Europe and China.

Shaking the Global Money Tree: International Finance

My friend Dom Phillips travelled there in 2019 as a jumping-off point for a reporting trip on illegal deforestation along the BR-163 highway that connects Itaituba to Brazil's agricultural heartland. 'It was striking to see how farming had eaten into the forest on both sides of the BR-163. Cows were everywhere. Wildlife survived as best it could,' Dom recounted in *The Guardian*. 'One morning, a white monkey scuttled across a dirt road, followed by a gaggle of forest pigs. Black, blue and orange macaws squawked atop a charred tree trunk, their only perch in a field of cattle. An opportunistic anteater darted across the highway in a gap between the trucks,' he added.

The Amazon's soy and cattle barons tell a cheerier tale, arguing that they have succeeded in turning the forest into the economic engine of a modern nation. 'Agro is tech, agro is pop, agro is everything,' as a well-recognised, long-running ad campaign produced by Globo, Brazil's leading broadcaster, puts it.

However, 'the numbers don't show that,' as Dom wrote in his notes for this book. 'Big Agro' is responsible for the lion's share of Amazonian destruction, having felled two Californias-worth of forest since Ford's death in 1947 and degraded an additional area of similar size, as it produces three-quarters of Brazil's greenhouse gas emissions. Meanwhile, multiple studies have demonstrated that major agribusiness operations actually tend to impoverish local communities and dramatically increase inequality.

The economic model still largely exists to serve foreign interests. And, now as then, it is underpinned by gunmen, land theft, slave labour, and a mindset that sees the forest and its inhabitants as enemies to be conquered with brute force. The great majority of farming and mining production, along with around a quarter of cattle, is exported. These activities are reshaping the region with government backing and the help of billions of dollars in investments and loans annually from foreign banks, corporations and investors.

But scientists warn that the forest is screaming for help and showing signs that its collapse may be imminent if we continue on this course, which would be catastrophic for many reasons, including imperilling the region's agriculture. A 2023 World Bank report suggests that the economic impact of Amazon destruction could reach US$317 billion annually, seven times greater than the profits that agriculture, mining and logging currently bring in. The worldwide climate impact could be far worse, considering that the carbon sequestered by the forest is greater than five years' worth of global greenhouse gas emissions.

'Man's conquest of Nature turns out, in the moment of its consummation, to be Nature's conquest of Man,' wrote British author and Ford's contemporary C. S. Lewis in his 1943 work *The Abolition of Man*. 'Every victory we seemed to win has led us, step by step, to this conclusion. All Nature's apparent reverses have been but tactical withdrawals. We thought we were beating her back when she was luring us on. What looked to us like hands held up in surrender was really the opening of arms to enfold us for ever.'

To avoid this dire future, the World Resources Institute's New Economy for the Brazilian Amazon proposal estimates that an additional US$516 billion in public and private investment is needed above baseline models by 2050, or about US$20 billion a year (0.8 per cent of annual GDP). They argue that would not only be enough to save the Amazon, but grow the regional GDP, improve social inclusion, add 81 million hectares of standing forest and reduce national emissions by 94 per cent.

Shaking the Global Money Tree: International Finance

It is simultaneously a colossal sum of money and a no-brainer investment, considering the cost of inaction, and not just for Brazil. Wealthy foreign investors, governments and customers who have never heard of a *jararaca*, *pororoca* or *pirapitinga* bear an enormous responsibility for bringing the Amazon to this deteriorated state. In global climate negotiations, they are under increasing pressure to not only fix their ways, but substantially fund the transition to sustainability for the Amazon and the rest of the Global South. But the people most responsible for the damage and with the deepest pockets to fix it have largely been slow to voluntarily cough up the cash or enact sustainable practices that contradict their short-term self-interests. For the forest to survive, that must change, and development in the region must prioritise local and long-term concerns. With new signals of a looming collapse appearing every day, already impacting supply chains and profits, some are beginning to realise that short- and long-term interests are converging.

In this sense, the Amazon's future is not being decided in the Amazon, or in Brazil for that matter. The region's conundrum is deeply enmeshed in the climatic, political, financial and social crises bearing down on the entire planet. Many of the most powerful potential solutions to its local issues are actually part of much larger global debates, like tax and subsidy reform.

So what is the best path forward? And do we have time left to experiment with incremental reforms or are more revolutionary measures required?

Starting in 2004, Brazil cut deforestation by 84 per cent in only a few years, proving that it was possible to overcome the greed, prejudice, ignorance, poverty and political failures driving the crisis. The dramatic setbacks during Jair Bolsonaro's presidency, however, exposed how fragile such gains can be without more systemic – and international – socioeconomic changes.

Dom had little patience for positions more inspired by leftist orthodoxies or paternalistic fantasies of a pristine forest inhabited only by mystical Indigenous heroes than actual praxis. Nearly half of the Amazon's 29 million residents are surviving on only US$66 per month. 'Who are we to judge them for taking the only opportunity in front of them to feed their kids, even if it is clearcutting to raise cattle or joining a gang of illegal miners?' Dom asked me more than once as we discussed this book project over beers.

That specific sentiment rattled around my head for months after Dom was murdered by people who, arguably, fit that description. The tragic irony of it all is that he was out in the Javari Valley that day trying to help the world empathise with people like his assailants, not criminalise or judge them. Their families deserve safety, dignity and a decent future; any realistic proposals, therefore, cannot simply abolish commerce, nor allow it to continue its current course.

Which is why Dom included a chapter on solutions from international finance and business in the proposal for this book. In discussions about how to complete this chapter for him, it was decided to broaden the scope a bit to also include foreign government action, as public and private policy are intimately related, as Dom himself may have eventually done. 'People can snub this and say capitalism will never have answers and in that they are right, but people also do need to make a living,' he wrote in his scant notes for this chapter. He included some of these notes in the chapter outline he sent to his publisher but, in private, he told me that they were mostly just a placeholder – this was the chapter he felt the shakiest on.

It is clear that, despite all the devastation that he had personally witnessed, Dom was an optimist about society's capacity to wield public pressure to force governments and corporations to do better. It was the driving motivation behind this book, after all. 'This mobilisation has already begun, though it has yet to produce results,' Dom wrote in the outline. Whether it will happen in time to avert disaster remains the question.

That is why, to advance his investigation, I travelled to the frontlines of the mobilisation – not in the Javari Valley, but 14,000 kilometres away, in the Arabian desert, where at COP28, Amazon environmentalists were squaring off against big agriculture, mining and oil lobbyists over the future of the Amazon and the framing of the global climate debate.

The smell of freshly brewed coffee wafted through the Brazil pavilion, which was buzzing with excitement. The atmosphere felt more like a music festival than a gathering of policy wonks.

Over 3,000 Brazilians had flown to Dubai, United Arab Emirates, for the United Nations' twenty-eighth Conference of the Parties, the most important global climate summit. COP is the main forum where international agreements to avoid (or guarantee) a climate apocalypse are struck, alongside immense amounts of networking, lobbying, pitching and preening.

The Brazilians were the biggest delegation at the event after the host nation. All of the government's heavy hitters were there, making major announcements like the launch of the 'Tropical Forests Forever' fund, a proposed mechanism to finance forest preservation with US$250 billion from wealthy nations and polluting industries. Every non-profit with a climate agenda was also in attendance, along with influential bankers and plenty of major polluters, hoping to score PR points and avoid going viral in any unscripted interactions with scrappy protestors. This was the first COP since internationally respected statesman Luiz Inácio Lula da Silva had returned to the presidency, replacing the far-right, climate change-denying Jair Bolsonaro. Lula was elected on a platform to save the Amazon, stop the devastation of Indigenous territories and kickstart the economy that had been flailing under Bolsonaro's chaotic management. But a large chunk of his shaky congressional coalition was made up of former Bolsonaro allies, who had other plans.

In just two short years, Lula would be presiding over COP30 in Belém, Pará, at the mouth of the Amazon River. 'Don't expect anything close to this!' Lula joked in one meeting, referring to the sprawling, opulent, seven-billion-dollar ExpoCity venue, full of Instagrammable modern art sculptures, glitzy architecture and a solar park, with its own metro station. 'If we can't offer buildings as fabulous as these, we'll give people trees to have meetings under,' he quipped.

Brazilian climate activists, still shellshocked from four years of brutal defeats, were daring to dream again and scrambling to piece together a cohesive, unified message. The Lula government, they hoped, would spend the next two years scoring climate victories and arrive at COP30, on the tenth anniversary of the Paris Climate Accords, armed with Lula's renowned diplomatic skill and fresh climate credibility to usher in a new paradigm led by Global South nations. To many of them, it was nothing less than humanity's last hope to save itself from . . . itself. And, depending on whom you spoke with, the path forward was either incredibly simple or maddeningly complex.

A visibly irritated panel moderator at the Brazil pavilion was again scolding the throng jammed into the lobby, whose loud chatter was drowning out one of the dozens of discussions about green finance, carbon credits, bioeconomy and the like. Starstruck fans excitedly asked for selfies with renowned climate scientists like Carlos Nobre and Brazil's first ever Indigenous cabinet minister, Sonia Guajajara, whom Dom referred to in private as 'one of the most important and inspiring political figures in the country'.

I found Guajajara out in the pavilion's sterile, cinderblock hallway posing for a photo op with another pioneering Indigenous politician, Joenia Wapichana. The gathered crowd, cell phones in hand, laughed as Wapichana stretched on her tiptoes to mark her hometown in Brazil's northernmost state of Roraima on a giant wall map – like so many things at COP, her objective was in sight, but just beyond reach. She let her serious public demeanour crack for a moment as she chuckled along with them.

Dressed simply in a red floral headband with a thin red line of paint across her cheeks and another down her chin, Wapichana wore an olive green T-shirt with 'National Foundation of Indigenous Peoples' written on the sleeve in Portuguese. The 50-year-old lawyer's diminutive height and casual apparel belied her immense stature in Brazilian politics. She was the first ever Indigenous woman to run this government agency, just as she was the first to be elected to Congress in 2018.

Guajajara and Wapichana, both from the Amazon, symbolized the Brazil that the people gathered around them wanted to believe in – a country that would finally put environmental protection and

Indigenous rights before profits and, in so doing, become a global pioneer in a new green political economy.

But it wouldn't be easy. Among the 85,000 international attendees at COP28 were a record number of corporate lobbyists from polluting industries, along with 34 billionaires and dozens of heads of state and royals, many of whom flew in on fleets of private jets and appeared more concerned with curbing policy proposals that could threaten their vested economic interests.

COP is 'a game of who can have more influence', Wapichana told me as we walked between buildings to her next scheduled appearance. 'Just as they point to us as an obstacle to development, it's important that we also hit their soft spots and point out that these business ventures are causing harm not only to Indigenous peoples, but to everyone,' she said.

Over her political career, Wapichana had seen the Indigenous movement make material gains from its engagement in international forums. 'Just the fact of being here, expressing our interests on climate, is already a step forward,' she said. But: 'we don't have much time, so everyone has to give up some of their beliefs.'

Not everyone was as optimistic. The main focus of the conference was on humanity's unsustainable overdependence on oil. An authoritarian petro-monarchy famous for money laundering, ultra-luxury tourism, abusive treatment of migrant labourers and artificial archipelagos shaped like palm trees was, therefore, a curious choice of venue to negotiate a fair and just transition from fossil fuels. But not nearly as curious as selecting Sultan al Jaber, the CEO of Abu Dhabi's billion-barrel-a-year national oil company, ADNOC, as COP28 president. One of the most polluting companies from the world's most polluting industry was not just 'getting a seat at the table', as oil execs had long demanded, the very expensive table was his.

Al Jaber, echoing Brazil's agribusiness and mining lobbies, argued that the only way to make a meaningful deal was to work with industries, but his credibility was undercut when the Centre for Climate Reporting revealed internal documents showing how he was leveraging his position to help ADNOC negotiate deals to significantly expand its oil and gas drilling, including its bid to buy a major stake in Braskem, the Brazilian petrochemical giant.

Projections suggest Dubai, the most successful of the seven United Arab Emirates, may become unliveable by the end of the century as temperatures rise – meaning the Emiratis have just as much to lose as the rest of us. Al Jaber and Wapichana's fates were inextricably linked in more ways than one.

Panic flashed in the eyes of venue security guards as a crush of Brazilians tried to barge, bargain and beg their way into the standing room-only event. An anxious aide stifled her irritation to explain to the guard that the grey-haired man perspiring in a suit next to her was the mayor of Belém, a Very Important Person, who was definitely on the list and needed to be let in. The exasperated guard winced and shrugged. He had no list.

The lucky ones who made it in injected the generic meeting room with vitality. They draped bright banners over the rectangle of grey tables, boldly announcing the presence of their coalitions. Sprinkled among the standard business attire of bureaucrats and bodyguards were environmentalists in neon tank tops and students in their organisations' T-shirts and turbans with joyous African prints. Indigenous leaders added to the mix in vibrant headdresses made with long red and blue macaw feathers that, all around the COP complex, drew curious stares and not-so-discreet cellphone photos from passersby.

Lula, flanked by his wife and ministers, began day three of COP sitting down with 135 civil society representatives packed into a small conference room. Such a meeting would have been unthinkable under Bolsonaro. It was meant as a signal of his commitment to environmentalist campaign promises made to his base. But, after it was over, all anyone would talk about was the big announcement that Lula *almost* made.

'Lots of people were alarmed that Brazil was going to join OPEC. We are not going to join OPEC,' the 78-year-old grumbled at the group, referring to the powerful cartel of major oil-producing nations leading the opposition to a fossil fuel phase-out.

On the COP's opening day, his minister of mines and energy had made headlines with an unusually timed announcement to the contrary. A beat of confusion passed. *Was Lula really taking a stand against fossil fuels? History in the making!* Some in the room began to applaud.

But he hadn't finished and cut back in quickly before the applause picked up steam. 'Brazil is going to participate in something called OPEC+! This name is so chic. OPEC+!' The air was sucked out of the room. He went on to say something about making change from the inside.

Outside, a visibly disappointed Marcio Astrini, executive director of the Climate Observatory, a coalition of 90 environmental organisations, huddled with colleagues. The timing of the minister's OPEC+ announcement caused 'disgust and revulsion' among many Brazil delegates and administration insiders, according to Astrini. It was 'intentional to sabotage everything Brazil had planned for that conference', putting Lula on the back foot and exposing the government's competing allegiances to dirty business lobbies and its mandate to avert a looming environmental catastrophe.

'Our big desire in that meeting was for Lula to become the main protagonist of the biggest item on the climate agenda, the end of fossil fuels,' remembered Astrini. 'The world needs a leader to move us in that direction.' Instead, Brazil earned the ignominious 'Fossil of the Day' anti-prize from Climate Action Network, in large part due to Lula's support for opening up vast deep-water oil reserves at the mouth of the Amazon river, just downstream from the future home of COP30.

On the same day as Lula's meeting, Colombia, Brazil's Amazonian neighbour and Latin America's third largest oil producer, became the first hydrocarbon producer to sign the Fossil Fuel Non-Proliferation Treaty, a pledge to ban new fossil infrastructure and phase out dirty energy in an equitable manner. The juxtaposition could not have been starker.

It was always going to be a major challenge to bring Brazil's most polluting industries – forestry, cattle ranching, factory farming, mining, transportation and oil – to voluntarily embrace ambitious climate goals that could cost them money in the short term, but their influence was evidently so strong that Lula could not even get his own team to stay on script for a few days as the world's eyes looked to them for signs of hope.

The oil drama overshadowed good news about Brazil's 50 per cent reduction in Amazon deforestation that year as well as the ambitious and intriguing 'Tropical Forests Forever' fund Brazil had announced the day earlier.

It is broadly accepted – in theory, if not in practice – that wealthy Global North nations must compensate the Global South for the historical environmental damage they have wrought, but those nations have been slow to contribute the cash. At the previous COP, a 'Loss and Damage' fund was announced to serve this purpose. The UN estimates the immediate need for US$215 to US$387 billion annually to offset climate damage, but, as of January 2024 only US$661 *million* in donations had been pledged.

One of the brightest new ideas to attain some of the necessary large cash transfers came not from the big banks or corporate lobbies, but Brazil's Environment Ministry under Marina Silva, a former rubber-tapper from the Amazonian state of Acre. Under the incipient Brazilian proposal, which should be exciting even to hardened sceptics, the fund would function essentially as a AAA-rated government bond, not a donation. The hefty US$250 billion fundraising target looks less daunting when compared to the global green, social and sustainability-linked bond markets' US$4 trillion in issuances in 2021 alone Managers would reinvest the money and pay out a fixed, market-rate return, with the remaining net profits being distributed to satellite-verified forest protectors.

The residents and landowners of 80 nations with tropical forests would receive money for each hectare of forest preserved, reportedly starting at US$25. A deduction 100 times greater taken for each hectare cleared makes the maths work at current deforestation rates and makes it more profitable to keep the Amazon standing. At that valuation, Brazil's largest Indigenous territory, the Yanomami, would receive a staggering US$241 million for a year of zero deforestation, nearly as much as the entire 2024 budget of the Ministries for Indigenous Peoples, Women and Racial Equality combined. And that is for less than 9 per cent of all Indigenous territory in the Amazon. Climate experts generally lauded the proposed framework for leveraging existing financial systems to deliver a clean and elegant solution to incentivise conservation that did not rely on philanthropy, complex new marketplaces, hard-to-verify CO_2 reduction claims, or allow contributions to serve as a green light to continue polluting. And, importantly, it does not require enormous shifts in geopolitical paradigms or new taxes and regulations.

The model's core control mechanism is similar to another Brazilian innovation, the successful – and considerably smaller – Amazon Fund, which releases money to be spent on forest preservation projects only if the country tangibly maintains its commitment to cut deforestation. Under Bolsonaro, the fund was suspended and Dom would be happy to hear that it has now been restarted. The fund follows the UN's Reducing Emissions from Deforestation and Forest Degradation standard, or REDD+, by which Global South forest communities and countries are paid by Global North nations. REDD+ mechanisms can be implemented philanthropically or used as the basis for carbon emissions offset schemes.

The fund launched in 2008 with a US$1.2 billion commitment from Norway and much smaller donations from others. It has dispersed about half of that amount to 107 projects under the management of Brazil's national development bank, or BNDES. The initiative has been the single most important financial lifeline for Amazon protection, albeit still far less than what is needed and with room for improvement. The majority has gone to government entities, much of it for law enforcement measures, which are at the heart of Brazil's deforestation strategy. Non-profits working on forest management, scientific research, Indigenous community support and sustainable supply chains have also benefited.

While representatives of the Amazon Fund and BNDES were well represented in the corridors of COP, notably absent were the small-scale farmers desperate for the kind of assistance that the fund could provide to help them transition to sustainable practices. They couldn't afford the trip.

Back on the agricultural frontier of southern Pará, Giselda Pereira, a 49-year-old farmer, called me from a dimly lit room. It was the first place she could get an internet connection on the road between her farm and an event she attended for female small farmers in the area. 'Very little' from the non-profit-run projects funded by the Amazon Fund or others 'has arrived out in the field to change our reality', explained Pereira with the measured confidence of a seasoned organiser. She is part of the national leadership of the Landless Workers' Movement, known as MST, which has settled 450,000 families on farms and, in the last four years, has gone a quarter of the way to its goal of planting 100 million trees.

In 1999, her collective occupied a large, unproductive cattle ranch in Pará's agricultural frontier. Eventually, after nearly a decade of expulsions and conflict, the government titled their plots. Most of the farmers continued to use the land for cattle for milk and meat, she said, because it was the cheapest option on land that they did not formally own. Lately, though, climate change has nudged them to replace pastures with more profitable and sustainable native crops, like manioc, açaí, cacao, and cupuaçu.

Pereira's collective and others in the area want to grow their agroforestry operations but lack the capital and equipment to restore and reforest the land. 'Agricultural collectives are not able to meet the requirements to even present a proposal [to the Amazon Fund]' – which has earned a reputation among potential recipients as being overly bureaucratic. 'Instead, large NGOs win the grants,' laments Pereira. Little other philanthropic money seeps through either.

Pereira wants to see lots of new funding to help families occupy big farms on illegally deforested land and convert them into sustainable agroforestry operations that increase forest cover. To the chagrin of many conservative political pundits, socialist MST has even dipped its toes into the Brazilian financial markets in search of capital, successfully issuing US$4 million in low-interest bonds and, to date, has made every payment on time.

BNDES' socio-environmental director Tereza Campello is a big believer in family farms but wants her bank to think bigger. 'Some of the projects are wonderful, but we have an assessment that [the Amazon Fund] ended up working on projects that weren't at the scale needed.' They had solid local benefits but 'did not lead to a change in the status quo' at a systemic level. That is to say, she wants to do more than just convert a few hectares of pastureland to agroforests.

Now, BNDES aims to work more closely with all levels of government to 'build agendas anchored in major public policies' focused on multiplying their socio-environmental impact locally. This emphasis, rather than foreign markets, is a significant shift, as much of the hype around a new Amazonian economy, sometimes referred to as the 'bioeconomy', has long focused on turning small-scale forest producers into competitive export powerhouses. But that has not been easy. 'People have been trying to organise the supply chain for 30 years. They can't,' said Campello with a bluntness uncharacteristic of a high-ranking bank officer.

Even the development of a growing foreign market for açaí, which is held up as a model of success, only represents a few tens of millions of dollars – more than 2,000 times smaller than Brazil's soy exports. The little purple berry rots quickly in the Amazonian heat and cannot be processed by small-scale producers, meaning they need to be shipped downriver to a processing plant in Belém as soon as they are picked. The numbers are not much different for another notable product, Brazil nuts, which have their own logistical challenges. The scale of these industries is not enough to make bankers in New York and London salivate, but BNDES' Campello is excited. She sees that stimulating the sustainable production of local crops and connecting them to local buyers, like public school systems, may not increase the trade surplus, but it can do a great deal to feed and strengthen Amazonian communities like Pereira's while increasing their self-sufficiency. And that is all that many of them are asking for.

Campello's colleagues are again raising philanthropic contributions from wealthy nations to expand the fund. Since Lula's return to office, European nations have pledged US$223 million, while an additional half a billion dollars promised by then-US president Joe Biden stalled in the Republican-controlled Congress, with only one-tenth of that amount released so far. The fund is still overwhelmingly financially supported by the Norwegian government's International Climate and Forest Initiative, or NICFI, which announced another US$50 million pledge at COP28.

On a park bench outside the Norway pavilion, sipping a cup of coffee and wearing a handsome, slim-cut grey suit, NICFI's director,

Andreas Dahl-Jorgensen, told me that, beyond funding REDD+ projects, they are 'working with the global commodity markets and financial markets to shift capital flows from deforestation activities' in a variety of ways. That means financing free access to satellite supply chain monitoring tools, developing forums where major deforesters and other stakeholders discuss solutions to clean up their supply chains, and investments in cleaner alternative business models, among other initiatives. But 'the most decisive factor', he is adamant, 'is government policy'. NICFI donates US$5 per tonne of carbon emissions avoided by keeping trees standing. The investment is purely philanthropic; it does not convert into carbon credits to reduce Norway's net emissions – but it does buy them some much-needed goodwill.

The Scandinavian nation's philanthropy is dwarfed by the billions of dollars in profits it earns annually in Brazil, mainly from polluting industries and the banks most responsible for funding deforestation, drawing heaps of criticism from climate organisations. Equinor, the state oil company of Norway, which is the largest oil and gas producer in Western Europe, plans to expand its global drilling in 2024 and quintuple its oil output in Brazil over the next decade; while state-owned Norsk Hydro brought in US$4 billion in revenue in 2023 from its Brazilian operations alone, including a bauxite mine and a smelter. Its aluminium refinery near Barcarena, Pará, the world's largest, is facing international legal actions for allegedly spilling toxic waste through a 'clandestine pipeline' and causing serious health problems, including birth defects, in downstream communities. Indigenous and *quilombola* communities are fighting to block a new ore pipeline running through their territories.

And yet, at the same time, Norway is legitimately a global climate leader in many of its other policies, like strong domestic carbon taxes and using its sovereign wealth fund to push companies to improve on climate performance.

Hundreds of oversized bronze punch cards strung together formed a mechanised simulacrum of a forest canopy that shielded Dinamam Tuxá from the harsh midday desert sun in Dubai's ExpoCity. The ruddy-faced lawyer from the Tuxá people looked sleepy as he lounged on a bench silently with a couple of colleagues. His black t-shirt with the logo of the Articulation of Indigenous Peoples of Brazil, the premier national Indigenous coalition that he leads as executive coordinator, was splotched with puddles of sweat and his hawk-feather headdress drooped.

As my voice recorder powered up, so did Tuxá, as if suddenly struck by a bolt of electricity. Staccato, razor-sharp insights sliced through the heavy mid-afternoon air, his passion and charisma contrasting with the subdued murmur of besuited Europeans networking over sandwiches, coffee and cigarettes on the benches around us. 'This is a global civilisational crisis,' said Tuxá. 'No measures, literally none, have yet been effective in controlling the violence against Indigenous peoples and deforestation. Just the opposite.' One in five environmental activists killed worldwide in 2022 lost their lives in the Amazon.

If you had to sort the messy jumble of the Save the Amazon debate into a single spectrum, the extremes would be 'the market will save us' and 'the market is killing us'. Tuxá falls squarely into the latter category. 'We're seeing agreements being negotiated that will worsen socio-environmental conflicts, like the one being discussed between Mercosur and the European Union.'

Shaking the Global Money Tree: International Finance

A major priority for Lula, a deal was signed in 2019 and a finalized version was agreed upon in December 2024, but must still be ratified by the EU and Mercosur nations. Bilateral trade agreements, said Tuxá, increase demand, which inevitably increase violations against traditional communities. These agreements 'do not look after people, they do not look after the forest, they are very focused on capital. It always comes first.' He felt the same way about the COP negotiations.

Brazilian land has quickly doubled, then tripled in value in recent years, as foreign investment has been used to buy up tracts the size of Belgium and rising foreign demand for agricultural products has pumped dollars into the industries most responsible for abuses. The Bolsonaro government accelerated this tendency by making it easier for foreigners to buy land and for farmers to get loans from abroad. His government also launched new securities for retail investors to buy into the frenzy. The result is a concentration of farmland in the hands of fewer and fewer highly capitalised buyers. The top 0.04 per cent of Brazilian farms – about 2,400 in all – are larger than the 4.1 million smallest farms combined, according to the most recent census data from 2017. The financialised, commodity export-focused trend has even led experts to raise alarms about Brazil's ability to feed itself in the near future.

Despite claims by agribusiness corporations and their financiers that they are committed to ethical and sustainable economic growth, watchdog Forests & Finance, which tracks financial institutions that fund big agro, finds their environmental, social and governance, or ESG, policies, 'woefully inadequate'. ESG is an acronym that has entered almost every corporation's lexicon, starting in 2004, as part of a movement that asks businesses to consider their societal impact and look beyond their bottom line.

In 2023, Forests & Finance reported that top investors scored only 17 per cent on average by their ESG metric. BlackRock, Vanguard and State Street – three of the largest global asset managers – were among the worst, joined by Brazil's government-backed Banco do Brasil, Banco da Amazônia (the only organisation to score 0 per cent) and Banco do Nordeste do Brasil, along with private Itaú Unibanco and Bradesco. Only four top creditors and investors broke 40 per cent. Most troublingly, the worst offenders had increased their involvement in forest-risk companies since 2016 – in the case of Itaú by 1,200 per cent.

Sitting at a metal café table on one of the main COP promenades late in the afternoon, I asked Forests & Finance's soft-spoken Dutch coordinator Merel van der Mark if there is another way that is both sustainable and profitable. Behind us was a cheery pink food truck selling pricey vegan Neat Burgers, a brand launched by celebrity environmentalists Leonardo DiCaprio and Lewis Hamilton.

'There are definitely models that can feed the world, make communities thrive and protect the forest. Are they profitable for BlackRock? Probably not,' she said flatly, pausing for dramatic effect and looking at me with blue eyes that reminded me of Dom's.

'It just doesn't make sense to produce beef in the Amazon. The opportunity costs are way too high,' added van der Mark, who spent many years in the region on the frontlines of land conflicts. The Brazilian government and many of the more corporate-friendly NGOs disagree – they favour investing in intensification to produce more cattle on less land. But for van der Mark, the issue is not efficiency, it's that 'there's a lack of political will' to be responsible. 'The fact that there's still no traceability is just ludicrous,' she lamented.

Tuxá agreed and was in Dubai to tell the world that the only way to stop their Brazilian business partners killing the Amazon's most effective defenders is with urgently needed foreign legislation that obliges companies to clean up their supply chains, going above and beyond Brazilian laws written by the agro lobby. He was advocating at COP for a bold solution that is favoured by some progressive climate activists abroad but unlikely to materialise soon: repurpose legally binding multilateral trade bodies like the World Trade Organization.

Historically, these organisations have been used to increase corporations' power and limited states' autonomy to rein in predatory multinational corporations operating within their borders. Tuxá wanted to retool these bodies with international enforcement authority – that is to say, the power to make companies follow the rules and actually punish them when they don't, not just issue a stern reprimand – in order to ensure rigorous and mandatory global environmental and human rights standards. Additionally, member states could form a trading bloc that benefits nations actually moving towards a more sustainable and equitable economy and punishes those still stuck in last century's slash-and-burn economy.

'We need to see people arrested, reparations, criminal and civil penalties for the crime of ecocide' in order to end the 'overwhelming sense of impunity', said Tuxá, letting out a deep sigh.

There have already been small signs of progress here. After years of lobbying by the financial industry to prevent, delay and weaken government action, regulators in the USA, EU and UK enacted a slew of new sustainability standards and disclosure regulations that took effect in 2024. One measure that may potentially signal a sea change is the EU Green Bonds standard, which created a uniform standard for any issuer who wants to claim their bond is environmentally sustainable. Without a consistent, binding standard, any company could use their own criteria to claim their corporate debt was 'green', even if the debt would actually fund rainforest-destroying practices, as meatpacker JBS did when it raised US$3.2 billion in the US in 2021. A report by the non-profit Finance Watch determined that the EU had made 'unprecedented progress' but loopholes 'currently undermine the efficacy of the EU sustainable finance regulatory framework'. Lawmakers needed to resist industry backlash, the report said, and finalise transparency and governance requirements, set mandatory targets for sustainability risks and improve enforcement measures.

The EU's groundbreaking but hotly contested Deforestation Law, passed in mid-2023, deserves special attention. It requires importers to prove that cattle, cocoa, coffee, palm oil, soy, rubber, wood and derivative products are not linked to any post-2020 deforestation or degradation, and authorises penalties of up to 4 per cent of their EU revenues for infringement of this, potentially hundreds of millions or billions of dollars for major commodity traders. The law will hold Brazilian producers who export to the EU to much higher standards than local regulations do, which depend on self-reporting, which is much less reliable and easier to falsify. Only 2 per cent of the Brazilian herd currently meets the EU law's requirement of individually tagged and tracked cattle.

At least 17 per cent of the cattle and a quarter of the soy the EU buys from Brazil may come from deforested land, which suggests the law will cause shockwaves in industries largely unprepared for such rigorous enforcement measures. Importantly, it makes no distinction between legal and illegal deforestation, preventing Brazil's agro caucus from legislating their way out of the regulation.

Brazilian agribusiness associations have complained that the additional cost of tracing supply chains will be too onerous and even deforestation-free small and medium-sized producers may simply opt not to export to the EU rather than comply. But if the local government and other major importers follow the EU's lead and embrace stricter policies, skirting regulations or choosing to export to buyers who look the other way would no longer provide a big enough market, forcing major producers and their suppliers to improve their sustainability.

Lula criticised the law's unilateral approach, telling the EU president he preferred 'strategic partnerships' and 'mutual confidence'. Environmentalists, meanwhile, are pressing for the law to be expanded to include financial institutions and non-forest ecosystems during a scheduled 2025 revision.

Brazil's main trading partners, China and the US, are also taking steps towards stronger standards, but lag behind Europe. The US is particularly hindered by climate denial and obstructionism of its increasingly radical Republican Party. In 2025, Donald Trump burst back into the White House withdrawing the US from the Paris Agreement on climate change for a second time, stripping environmental regulations, and bolstering a right-wing backlash to ESG policies.

At the other end of the sprawling COP complex, I met up with Roberto Waack, who had a sunnier outlook. He speaks in measured, qualified assertions interspersed with rhetorical questions and long pauses to choose his words carefully. 'The rainforest economy is an invisible economy. The value of nature is still little recognised in the financial market.' This is both a lament and a business proposition.

Slim, bespectacled, and sweating through his blue button-down shirt, Waack sported a grey beard that gave him the appearance of a kindly biology professor. He acknowledged many of the same problems as Tuxá, but saw a friendlier face of Amazonian capitalism after a career spent as an executive in the forestry, paper and pharmaceuticals industries, including as CEO of a certified B corporation, a designation for businesses that meet certain standards for social responsibility. Waack also sat on the board of prominent environmental non-profit organisations and Marfrig Global Foods, which, after JBS, is the second largest beef producer in the world and has repeatedly been caught with illegal deforestation and slave labour in its Amazonian supply chain.

Some environmentalists might see those credentials as an inherent contradiction, but not in Waack's worldview. Even an entirely green and sustainable economy is going to need industry and lots of *stuff*, and Waack thought we should work from within and do what we can now for our food, houses and daily essentials to be produced more ethically. Rather than viewing big business as inherently criminal, he saw criminality as the biggest threat to those businesses. Governments, civil society, large corporations and international investors should work in coordination to create 'structuring actions' that will shape and solidify the regional economy, said Waack.

This means, essentially, incentivising large corporations like Marfrig to invest and then rely on pressure from consumers to convince them to act with social responsibility. The government's role is mainly to stay out of their way and help create a financial ecosystem that will create trickle-down benefits to smaller companies in their orbit, that will, in turn, stimulate the local economy. No new regulations necessary.

'The business environment in the Amazon has to be less bureaucratic, with society keeping a close eye on the products generated by these companies,' explained Waack. But don't we pay taxes because the government is supposed to do that on behalf of society?

'The way I see the hyper-regulatory process, it ends up being inefficient because the companies that really have good socio-environmental programmes see [regulations] in a negative light, in the sense of transaction costs, and flee the region,' countered Waack, 'leaving only those companies that somehow manage to escape from the regulatory system in heterodox ways' – which is Waack's way of saying 'commit crimes'. Regulations 'work in environments where you have effective good governance'. But that is not the case in the Amazon and won't be any time soon, he argued.

Instead, Waack favoured models like social licences to operate, or SLOs – voluntary measures that companies take to achieve ongoing acceptance from local communities impacted by their operations and other 'stakeholders'. They can include commitments to share some of the benefits and compensate for damages.

In Juruti, Pará, US aluminium giant Alcoa was able to start building a bauxite mining facility in 2005 over the objections of the Public Prosecutor's Office and locals, who alleged they were promised a sustainable partnership that never materialised. That is, not until 2009, when 1,500 community activists occupied the port, highway, railway and mine entrance for nine days of protest, impeding Alcoa's operations. After legal challenges, police violence and attempts to split the community failed, the multinational was forced to the bargaining table.

The community and Alcoa negotiated a formal agreement, mediated by the state's Public Ministry, to distribute a share of their profits and pay for loss and damage, like spills that contaminated land and water previously used to sustain the communities' livelihoods. Such agreements open the door for a more constructive and collaborative relationship that Alcoa refers to as an SLO. The arrangement is now viewed by many as a more socially conscious model to be emulated for large enterprises in the Amazon and elsewhere – but *before* the first shovel touches dirt.

In addition to local social infrastructure projects and US$16 million in direct payments to the local community association, the company has paid US$159 million in royalties and taxes to the city, state and federal governments – though it took 14 years to eventually agree on loss and damage payments. These sums come to around 0.1 per cent of Alcoa's revenue, which in 2023 was US$10.55 billion. Such negotiations – particularly when conducted informally – begin with an immense power imbalance between multinational corporations and often impoverished local communities, but constructive state involvement can help level the playing field.

Free, prior and informed consent from Indigenous and traditional communities to use their land is legally required by an International Labor Organization convention signed by Brazil, but it is frequently ignored. If the treaty obligations were followed, loss and damage and profit-sharing agreements could be considered minimum table stakes for companies to sit down for negotiations, since environmental impacts of even well-managed operations can easily deprive communities of their livelihoods and way of life. If multinationals are not prepared to share the wealth, maybe keeping it in the ground is for the best.

In an example of more democratic alternatives, in Ecuador, a working Amazonian oil drilling project was put up for a national referendum and blocked by voters in 2023. The government is now obliged to decommission the facility.

Voluntary measures like social licences to operate are examples of environmental, social and governance policies. Multiple elaborate ESG ratings mechanisms have been devised to measure how corporations are considering factors beyond next quarter's returns. But the terminology as it is often used on Wall Street is rather counterintuitive, and intentionally so. It does not in fact concern itself with the company's impact on the environment, local society or governance, but rather on those factors' potential impact on earnings in future quarters.

Henry Fernandez, CEO of Wall Street's leading ESG ratings firm MSCI, told Bloomberg News in 2021 that his unregulated ESG business 'is 100 per cent a defense of the free-enterprise, capitalistic system and has nothing to do with, you know, socialism or zealousness or any of that'. Bloomberg found that less than 1 per cent of MSCI's ESG ratings upgrades had to do with actual cuts in emissions. When the reporters asked Fernandez if most investors understood what his ratings actually measured, he responded, 'No, they for sure don't understand that.'

Since then, ratings agencies have come under increasing scrutiny and thousands of funds and investment firms have removed ESG labels from hundreds of funds, in part due to regulatory pressure.

The chasm between ESG hype and practice was brought into sharp relief in a viral 2022 presentation by Stuart Kirk, HSBC Asset Management's then-lead ESG executive, titled 'Investors Need Not Worry about Climate Risk'. Why? Because 'at a big bank like ours, at HSBC, what do people think the average loan length is? It's six years,' Kirk told the audience of bankers and journalists. 'What happens to the planet in year seven is actually irrelevant to our loan book.' HSBC, a major investor in Amazonian industries, had signed off on the presentation's title, theme and content, but quickly distanced itself after public backlash.

'It's the bullshit that's killing us and they've been spewing this bullshit for years,' Tariq Fancy told me. In 2018 he became the inaugural chief investment officer for sustainable investing at BlackRock, the world's largest asset manager. But he soon quit after being pressured to over-promise real-world impacts of their ESG products that were ambiguous at best. He realised that his colleagues were interested in 'sustainability' only insofar as the label helped make sales.

'I would be very sceptical of any solutions that are voluntary in nature. Because a lot of these things are not what companies want to do. It's not profitable, otherwise they'd already be doing it,' cautioned Fancy. 'You will not change the behaviour of the participants in the financial services industry until you change their incentives. ESG not only doesn't change incentives, it actually distracts us from the fact that we're not doing anything.'

Fancy's experience in the industry and conversations with other insiders led him to conclude that ESG is a 'placebo' meant to 'delay regulatory changes' and prevent legislation that 'makes the economics work' by making polluters pay for externalised costs, like emissions, that are cheaper to leave for society to cover. The industry will only move when new legislation is introduced that forces polluters to pay for the damage they cause, says Fancy.

In the last few years, regulators and climate activists, such as Mighty Earth, have successfully used US courts to target major Amazonian polluters like meatpacker JBS and mining giant Vale for fraudulent ESG claims and greenwashing. So far, the victories do not 'make the math work' enough to substantially change corporate behaviour, but it is a promising trend that could be hastened by more forceful legislation.

Deforestation laws, ESG standards and fines were not major risks that worried Kirk when he was at HSBC. The one prospect highlighted in his presentation as truly disruptive was 'a whopping great carbon tax out of the blue of 200 bucks. Wham! Put that in your model.' Carbon taxes raise the cost of emitting carbon into the atmosphere for consumers and corporations in an effort to incentivise less carbon-intensive behaviour and raise funds for greening the economy. Currently, 27 countries have some form of carbon tax, but only four charge over US$100 per ton of emissions. Studies vary widely, but some claim prices need to move closer to US$200 to substantially change behaviour at the scale needed. Carbon taxes, however, have proved politically difficult in many countries, the US included, due to corporate lobbying and also because, if implemented alone, they are a regressive tax on consumption, disproportionately affecting the poorest – a recipe for popular discontent.

But other tax solutions to fund the transition and disincentivise polluters are more palatable to voters – at least, to the ones who don't own private jets: tax the rich and corporations. The Lula administration has put forward a new tax on corporate dividends and a wealth tax to raise a combined US$30 billion a year – which would be enough to cover the World Resources Institute's half trillion-dollar goal for modernising the Amazonian economy with a third left over.

'We need to understand climate change and poverty as truly global challenges to be tackled through a new socio-environmental globalisation,' Brazilian Finance Minister Fernando Haddad said in a G20 meeting in 2024, during a pitch to other countries to also adopt a wealth tax to help tackle climate change. Three-quarters of G20 millionaires support higher taxes on their privileged cohort, who are also the biggest polluters.

Oxfam, which has labelled Brazil a 'tax haven for the wealthy', is among those calling to 'tax the rich to save the planet now'. It estimates that, globally, a wealth tax on the ultra-wealthy could bring in US$1.7 trillion a year and a 60 per cent income tax on the top 1 per cent is worth another another US$6.4 trillion – which would also result in a United Kingdom-worth of emissions reductions. A windfall corporate tax on skyrocketing profit margins could yield up to almost a trillion dollars.

At COP28, UN Secretary General António Guterres called for taxes on windfall profits of fossil fuel companies, which should go to bankroll loss and damage funds for Global South countries and to people struggling with rising food and energy prices. The same could apply to other harmful industries to fund the transition and make them less attractive to investors. Windfall taxes were passed in Italy, Spain and the UK, but lobbyists defanged the legislation with copious loopholes, according to watchdog Fossil Free Politics.

A historic global minimum corporate tax rate of 15 per cent, agreed upon by 140 nations in 2021 to combat tax avoidance, should have added US$270 billion in revenue, but, two years later, corporate lobbyists again carried the day and gaping loopholes were added to the policy.

Money, it seems, is not in short supply; it is just being allocated to serve powerful corporate interests, rather than for urgent needs, like saving the Amazon and fighting climate change. This was former US Vice President Al Gore's focus in his presentation to an auditorium full of climate policy nerds. The 75-year-old bounced around the stage in an oversized blue suit as he enthusiastically clicked through the latest of his famed PowerPoints.

'Here's something to put it in perspective,' Gore bellowed to the packed room. 'The amount of the pledges into the [UN's] Loss and Damage fund equals 45 minutes' worth of annual global government subsidies for fossil fuels [in 2022, valued at US$7 trillion]. Governments around the world, many of them, are forcing their taxpayers to subsidise the effort to destroy the future of humanity.'

Half of all discovered oil in the US, for example, would be unprofitable to extract without subsidies, according to the Stockholm Environmental Institute. In Brazil, direct fossil fuel subsidies amounted to US$222 billion over the past decade. These funds could instead be directed to develop sustainable industries, build vital infrastructure and increase social spending aimed at reducing inequality in the Amazon.

João Peres, mild-mannered and quick to flash his warm and easy smile, believes the same logic may need to be extended to the industries most responsible for deforestation and threatening communities and biodiversity: cattle, factory farming, paper and pulp, and mining – all of which receive subsidies. Peres, one of Brazil's leading food systems journalists, founded *O Joio e O Trigo*, an independent newsroom focused on the nation's most influential industries. He has reported on how Brazil's soy industry receives at least US$84 billion in known government assistance and loans – many times more than what is offered to small farmers like MST's Pereira – though the total cost is actually unclear.

'It's worrying that we don't yet have a study that allows us to understand the true total cost of these subsidies,' Peres told me. He said that foreign philanthropic funding is essential to make this research happen. 'We urgently need to catch up with the rest of the world on this important debate: whether de-financing agribusiness in order to build another socioeconomic paradigm is *an* option or *the only* option.'

Peres emphasises the fact that subsidies are just the tip of the iceberg in terms of the social costs of this model, and it goes far beyond deforestation. The UN's Food and Agriculture Organization estimates that industrial food systems already generate up to $12 trillion a year in hidden costs, particularly from health problems and associated productivity losses, nitrogen emissions and, in some countries, extreme poverty.

It is clear that taking on the trillions of dollars in corporate welfare and tax avoidance will take more than an Al Gore PowerPoint, not least because Brazilian politicians have proved to be uninterested in challenging powerful business interests. They are more willing, however, to consider climate proposals that create new profit opportunities, like market-based carbon trading.

In December 2023, Brazil's lower house of Congress approved a regulated carbon market that would impose mandatory limits on emissions from major industries, and provide a marketplace for companies who surpass their cap to buy credits and those who come in it to profit by selling their allotted credits. Thirteen countries, as well as the European Union, have implemented this kind of mandatory emissions trading system, commonly known as 'cap-and-trade'. The specifics can vary but, generally, the government determines a set amount of emissions nationally and for each company, which decreases each year. More bureaucratic than a carbon tax, cap-and-trade systems provide more margin for loopholes and opportunities for well-positioned wheelers and dealers to make profits without actually impacting the climate crisis.

A major loophole was included in the Brazilian bill, currently before the Senate: agribusiness, responsible for three-quarters of emissions and most Amazon destruction, would not be required to participate But they could still reap major benefits by selling offset credits to companies.

Currently, Brazil only has voluntary carbon markets, where companies and countries who want to show lower net emissions to meet pledges purchase 'carbon offsets' or 'carbon credits'. These credits come from carbon sequestration projects and nature-based solutions, be they an Indigenous community under a decades-long contract promising to preserve their forest or an industrial plant turning pig poop into biogas.

The Intercept Brasil, where I work, revealed how American 'carbon cowboys' have unscrupulously acquired the rights to large swathes of the Amazon – including some public land – to sell millions of dollars of carbon credits to foreign companies, keeping the great majority of the profits for themselves and provoking social conflict within *ribeirinho* and Indigenous communities.

Similar situations across the Amazon are evidence of how 'green finance, green economy and ESG' – widely touted at events like COP – actually 'end up legitimising a new phase of neocolonial expropriation', said Larissa Packer of GRAIN, a non-profit that supports small farmers and social movements. It is 'a legal framework that authorises the appropriation of what had been common, public goods and transforms them into private property traded on financial markets,' said Packer. Or, as the *Financial Times* called it, 'a land grab'.

HOW TO SAVE THE AMAZON

Giselda Pereira from MST told me that she has seen predatory offers to small farmers in her area, who were promised quick and easy cash, but handed 25-year contracts with extremely unfavourable terms, firm restrictions on land use that make it difficult to plant and harvest crops, and no guarantee of payment. She believes that, with government assistance and training, small-scale farmers working together can do more to reforest the Amazon and provide dignity to people than carbon credits.

Worse yet, carbon credits rarely work. *The Guardian* and the watchdog Corporate Accountability evaluated leading providers and determined that 96 per cent had 'one or more fundamental flaws' in their methodologies and at least three-quarters were 'likely junk'. Influential players – including Norway, which plans to purchase large amounts of carbon offset credits to reach net zero emissions by 2030 while increasing oil drilling – continue to support international carbon trading. They argue credits are an important mechanism to incentivise conservation and transfer the enormous sums of money that developing nations will require for the green transition.

By the end of my conversation with Dinamam Tuxá, sitting together on that park bench in Dubai's ExpoCity, surrounded by diplomats, scientists and lobbyists, his electric charisma was spent. He let out a deep sigh and looked around. 'As long as the people in charge have this mentality of capital before humanity, unfortunately, we will continue in this global crisis.'

He had the distant look on his face that I saw in so many people I spoke to as they contemplated the immense challenge ahead and the prospect of the end of the world if humanity continues on its current path.

Tuxá knows that our future is a collective one, that he needs the people at COP and vice versa. But how do you convince this swarm of gringo individualists who buzzed around us that what is needed to save the Amazon is not just this or that policy, but a completely different worldview? Can they ever comprehend what is at stake without experiencing first-hand the Amazon's wondrous power?

'We are tired of fighting to protect something that benefits everyone and that is still not recognised here,' Tuxá said solemnly, 'but we will not give up.'

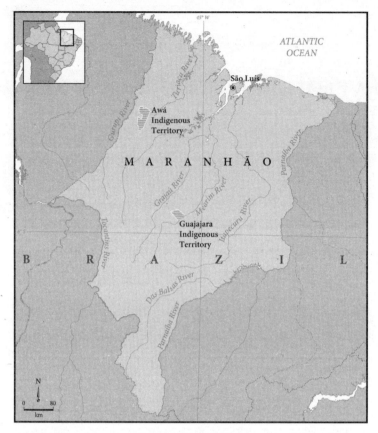

Maranhão State

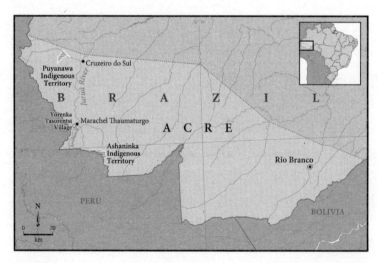

Acre State

CHAPTER NINE

Nature Worth Fighting For: Biopharmacy and Bioeconomy

Jon Lee Anderson and Dom Phillips
Ashaninka territory, Acre State, and
Awá territory, Maranhão State

Jon Lee Anderson is a staff writer for The New Yorker *magazine. He has reported from numerous countries and covered over two dozen conflicts since beginning his journalism career in 1979, in the Peruvian Amazon. His books include* Che: A Revolutionary Life, The Fall of Baghdad *and* Guerrillas: Journeys in the Insurgent World. *Anderson lives in Dorset, England, but returns to the Amazon, where it all started, whenever he can.*

'The world's largest library of biological knowledge.'

In the proposal for this book, Dom wrote the following:

In 2015, I was walking in a forest in the eastern Amazon state of Maranhão with members of the Awá people. Some 400 lived in villages, but an unknown number were still living isolated in the forest and occasionally appeared. The Awá villagers I had joined were going fishing, with perhaps a little hunting too, and took a break beside a muddy stream where one teenage boy scooped up a handful of foam from the top of a termite mound and smeared it on his head. As the boy gazed, increasingly fascinated, at an ant, I asked what this foam was for. 'To feel good,' one Awá man said and invited me to try it myself. 'It helps with hunting.'

I patted a much smaller amount of the foam on my head and soon began to feel altered. The feeling was not unpleasant, as the forest around me began to look and feel different. I found myself watching a small leaf perhaps ten metres in front of me as it slowly spiralled towards the ground. A spider spinning on a tiny thread dangling from a tree came into view, as if a camera had just zoomed in on it, like in a movie where the background moves out of focus. I could see how that ability to focus could help a hunter.

The effects wore off within an hour. But whatever the psychedelic termite foam was, I have never found a written reference to it and have only found one Indigenous man, Daniel Mayoruna, in the Javari Valley on the other side of the Amazon, who described something similar.

It was just a hint of how the Amazon's vast biodiversity is likely to contain drugs and remedies that the outside world knows very little about. One book by the University of Brasília has catalogued 450 Amazon plants with medicinal and alimentary uses for people and animals. In the Raposa Serra do Sol Indigenous land, located in savannah on the northeastern fringes of the

Nature Worth Fighting For: Biopharmacy and Bioeconomy

Brazilian Amazon, Indigenous health agent Leodora da Silva showed me shelves full of plastic pots of natural Indigenous remedies in the little wooden health post she manned in her community outside Uiramutã – a border town in remote hills reached by hours of treacherous dirt road. One paste she gave me healed cuts and helped mend the Caesarean scars of an acquaintance in Rio de Janeiro. Another potion cured fungal infections.

During the coronavirus pandemic, Indigenous people turned to their own remedies, making traditional teas and baths of leaves like *mastruz, jambu* and *folha de pirarucu* with garlic and ginger – and in some cases adding aspirin. 'We were raised with these traditional remedies, our parents brought us up with these leaves,' said Ednéia Teles, an Arapaso woman from the FOIRN Indigenous association in São Gabriel da Cachoeira, the Amazon town with Brazil's second biggest Indigenous population, which was hard hit by Covid-19.

The coronavirus pandemic highlighted the need for greater urgency in protecting the forest. Covid-19 is a zoonotic disease, crossing from animals to humans, and the number of new zoonotic viruses against which we have no immunity is rising. Land use changes play an important role in this. In 1999, the Nipah virus killed over 100 people in Malaysia after bats infected pigs. The bats had changed habitat after their forest was burnt and destroyed for industrial farming. A study from University College London in 2020 found that 'animals known to carry pathogens (disease-causing microorganisms) that can infect humans were more common in landscapes intensively used by people,' according to *Science Daily*. There is so much life in the Amazon that has yet to even be studied. And that immense biodiversity contains both threats and possibilities.

An estimated 80 per cent of the world's species are still undiscovered. How many of them are in the Amazon? Tapping the potential benefits of this vast reservoir of genetic and biological knowledge is one of the goals of synthetic biology – which has in recent years replaced petrochemical products and even given Burger King its vegan Impossible Whopper made of engineered plant protein. Algorithms developed from ant behaviour helped streamline logistics networks – Southwest Airlines used such technology to improve their cargo routing.

Juan Carlos Castilla-Rubio is a Peruvian entrepreneur and biochemist based in São Paulo whose company Space Time Ventures develops projects that use natural resources and nature's intelligence for global impact. Castilla-Rubio calls the Amazon rainforest 'the world's largest library of biological knowledge' and is part of the Earth BioGenome Project – a 'moonshot' which aims to 'sequence, catalogue and characterise the genomes of all of Earth's eukaryotic (that is, cells that have a nucleus) biodiversity over a period of ten years'. Castilla-Rubio is using blockchain technology to make Amazon biological knowledge visible and accessible, while keeping it controllable – avoiding the biopiracy issues that have meant previous discoveries did not benefit Amazon peoples. He believes that the Amazon rainforest could provide 20 per cent of antimicrobials needed to fight antimicrobial drug resistance – so-called superbugs. Frogs like the Amazonian giant monkey frog (*Phyllomedusa bicolor*) may, he said, have the 'bio-production instruction sets' in their DNA – its poison has long been used in Indigenous cleansing rituals and has become popular in alternative medicine circles.

The post-Covid-19 pandemic era offers the possibility for an opportune rethink about the way our modern societies operate. This can involve helping people understand how what they buy in a supermarket or restaurant and where they invest their money influences the climate conditions that they and their children will live under and has a bearing on future pandemics they will likely face. The Ashaninka Indigenous people in the western Amazon state of Acre, whose history stretches back to the Incas and whose sustainable agroforestry, rich culture and striped, flowing tunics and round straw hats with plumed feathers stand out in the region, have already laid down some useful potential pathways that are available to the rest of us.

Benki Piyãko is an Ashaninka leader, ambassador and agroforestry specialist who has travelled the world to explain how he and others replanted 3 million trees to reforest their region, created dozens of fishponds to restock empty rivers and worked to reintroduce wild animals to the forest at the Yorenka Tasorentsi Institute he set up to provide traditional and agroforestry education. Ahead of the curve, the community gave up cattle farming decades ago and reforested degraded pasture with fruit and coconut trees. Benki's brother Isaac Piyãko, a teacher, was re-elected as the mayor of the Marechal Thaumaturgo municipality where their reserve is located.

Benki and another brother, Moises, are both shamans. Their spirituality shapes their view of the forest and serves as a practical guide to our collective survival. 'We have to share responsibility for what we eat in our kitchens,' Benki told me in a video interview for the Oxford Real Farming Conference. 'We have to plant fruit, we have to plant forest, we have to take care of the rivers, we have to take care of ourselves.' Benki argued that his elders long warned of the destructive impacts of cutting down forests and argued the world needs to rethink its approach to combat the climate change and environmental destruction his people are already feeling, as the fish in their rivers die, the weather changes, and their forests are threatened by loggers and hunters.

'If nobody takes responsibility, the whole world will pay. Because nobody will have water to put fires out. Nobody will have water to drink. Nobody will have air to breathe any more. So we will have to step back a little and rethink our future, with our awareness and our humanitarian respect,' Benki said. In the cosmological view of the Ashaninka, he explained, everything on Earth is a cell that is part of the same organism, the same beating heart.

'We have to think of a different model,' he said. 'Everything is connected. Nothing is disconnected in this universe.'

Nature Worth Fighting For: Biopharmacy and Bioeconomy

I met Dom on 1 January 2019, in Brasilia, during the presidential inauguration of Jair Bolsonaro; I was on assignment for *The New Yorker*. We were introduced by a mutual friend, Carol Pires, a Brazilian journalist who was assisting me in my reporting. I had read Dom's work on the Amazon in *The Guardian* and admired what he was doing. We chatted and exchanged our mutual feelings of trepidation about the coming Bolsonaro presidency. We never met again but stayed in touch after that, usually corresponding via WhatsApp after one or the other had published something. The damaging effects of Bolsonaro's presidency, especially on the Amazon, was our major topic of discussion.

At one point, on 4 June 2019, I asked Dom: 'How dire is the Bolsonaro effect on the Amazon? Will he prove so incompetent that the damage will be mitigated, or is that going to hit anyway?'

Dom replied: 'Well, the damage to protection agencies etc. is scarily drastic and there are signs that this is already having an impact. My view is that this will prove disastrous because it's a green light to loggers, land grabbers, *garimpeiros* and so on in an already lawless area.'

In November of that year, I congratulated Dom on a piece I'd read of his about a trip he'd made into the Javari. He replied, clarifying that it was a previously published article he had written after his trip there in April 2018. 'I reshared, because the [Javari Indigenous reserve boundary protection] base was attacked again last week. Jesus. Just wooden huts on a river. A guy with a flashlight in a tower. Coming under shotgun fire in there must be scary.'

'Bastards,' I replied, and said I was looking forward to his forthcoming articles.

HOW TO SAVE THE AMAZON

When the news came that Dom and Bruno Pereira had gone missing on a trip to the Javari Valley in June 2022, I feared the worst, and when the news came that their bodies had been found I felt angry and sick at heart. I wrote an article for *The New Yorker* attributing their murders to what Dom had accurately described as Bolsonaro's 'green light' to lawless fortune-seekers in the Amazon. Later, when I was invited to join the other contributors to help complete Dom's book by tackling the chapter he had outlined on Acre, I readily agreed.

In March 2024, I finally made it there, travelling with my son Máximo, who, like me, cared about the Amazon, knew of Dom's work and who, a decade earlier, had spent the better part of a year living and working with Kanindé, an indigenous rights NGO in the neighbouring state of Rondônia. In our own effort to uncover the potential for an Amazonian bioeconomy, we decided to begin where Dom had left off, by paying a visit to Benki Piyãko, of the Ashaninka people, a personality in whom Dom had placed a great deal of hope for the future of the Amazon. I arranged to meet Benki as he paid a visit to the territory of the Puyanawa, another Indigenous group. One of Benki's sons, I learned, was about to be married to a daughter of a senior Puyanawa leader.

It was the day before the wedding ceremony and Benki was staying at the home of his in-laws-to-be. It was a large, rustic, two-storey wooden house built next to a pond in a cleared forest setting. Benki had an entourage with him that included some of his relatives, as well as acolytes from other countries, including Italy, Germany, the United States and India. These people evidently believed themselves to have been transformed by their experiences and thought of him as a kind of guru figure. Initially, most seemed

Nature Worth Fighting For: Biopharmacy and Bioeconomy

to assume I was there for the same reasons. I had not been at the house very long before a woman from New York, wearing a dress decorated with jungle motifs and her face and arms tattooed with jenipapo dye, approached me to ask solicitously: 'Is this your first *journey*?'

One of Benki's closest disciples was Federico Quitadamo, an Italian man in his late thirties, who told me he had given up a successful career as a private equity investor to devote himself to Benki's cause. Federico explained that he now lived in Acre himself and helped Benki to organise international retreats, in which he spoke about his efforts to save the Amazon and guided people through healing experiences using traditional medicine from the rainforest. These retreats were held in various parts of Europe, the US and the United Kingdom. 'Why does someone like me give up everything at the apex of his career to do this?' he said. 'Because I believe in what Benki is doing.'

To Federico, the undeniable proof of climate change meant that the destruction of the Amazon, and of the Earth itself, was inevitable without dramatic new policy initiatives and huge investments from the world's major governments and its corporations. 'The fact is is that as long as you work within the capitalist system, you can't save the Amazon,' he said. He saw Benki as a key inspirational figure in this campaign, while he used his own contacts in the world of international business and finance to gain the support of people of influence. 'The fact is it's not just about the tree you're planting, it's about what you're *starting* when you do that,' explained Quitadomo. When people in the financial world stop and listen to Benki and then start to help us, *that*'s when the big change begins.'

HOW TO SAVE THE AMAZON

Benki Piyãko, who is now in his fifties, became famous in Brazil as a young teenager when the legendary Brazilian singer and songwriter Milton Nascimento met him on a visit to his ancestral village and composed a hit song, called 'Benke', which he included in an album he released in 1992 in support of the Amazon and its Indigenous peoples. The scion of a family of Ashaninka *caciques* (leaders) whose father, a noted shaman, had married a woman of Italian descent, Benki was just a precocious boy of eleven when Nascimento met him, but was already regarded as something of a shamanistic prodigy.

It was a tumultuous time, with Brazil finding its feet as a democratic nation after its 21-year military dictatorship. The rights of Indigenous peoples had been recognised for the first time in a new Constitution and, across the country, lands were being legally demarcated and allocated to Indigenous communities as their inalienable territories. There was resistance to this movement, sometimes violent, from miners, loggers, settlers and ranchers whose activities had been encouraged by Brazil's military rulers. The westernmost state of Acre, bordering Peru, had become a frontline in the battle over the Amazon and had already produced some iconic figures, such as the rubber tapper, unionist and environmental activist Chico Mendes – murdered there in 1988 on the orders of a land grabber – and the activist and politician Marina Silva, Brazil's future environment minister, who, like Mendes, with whom she had worked closely, was also a rubber tapper's child.

In 1992, Benki's Ashaninka people were granted a vast 211,000-acre (85,400-hectare) reserve in the forests of the upper Juruá, one of the Amazon's longest tributaries. That same year, the first Earth Summit was convened in Rio de Janeiro. Benki, who was then 18, left his village for the first time to attend the Rio Summit. He travelled there by bus, and the 2,300-mile-journey had a profound impact on him. 'I saw a Brazil that was very destroyed,' he recalled. 'I learned that the laws of the land were not the same as the laws of the universe,' he told me. When he returned home, he saw the world differently. 'The forest is a house. A big house, one that you never get to the end of.' Soon, Benki had embarked upon his life's work – 'to protect the earth and life itself' by reforesting areas in Acre that had been cleared of trees. Using an oracular tone, he said: 'I feel like water and air, and I have this message for the people: if we kill the forest, we die too.'

When we spoke, Benki was lying in a hammock trying to shake off a fever that had overwhelmed him the day before. A slender man with an easy smile, Benki was shirtless and unadorned except for a pair of colourful beaded bands on each wrist.

I told Benki about Dom's unfinished book and the title he had chosen for it: *How to Save the Amazon*. Beki smiled sadly when he heard Dom's name, recalling their conversation, and expressed his sorrow over Dom's murder. I asked Benki what he understood by 'bioeconomy'. What did it mean to him and what part could it play in rescuing the rainforest from destruction?

'The only thing is to plant trees,' Benki replied immediately. 'That's it. Without that, we're not going to save the Earth. All the other forms of bioeconomy, really, are just thievery.' He explained that in his view, most carbon credit offset investments are essentially money-making schemes based on specious environmental claims, but he did support reforestation projects. He described his own Yorenka forest as a Garden of Eden, a place full of life. He had planted more than three million trees there, he said proudly, on over 1,600 acres of degraded wilderness land. 'It was all pasture then and now we have everything. Fish, fruit – it's all there.'

Despite the great calamity posed by the looming spectre of climate change, he went on, it was essential that people felt optimism, and with that in mind he was seeking to create a positive model at Yorenka that could help guarantee a sustainable future for people who lived in forest areas. Doing this was not easy. Beyond the usual threats from loggers and ranchers, Benki said he was worried by the proliferation of illegal narcotics and the outlaw culture that went with it. Acre is on the borders of Peru and Bolivia, both major producers of cocaine, and the drug was increasingly being smuggled through the Amazon to the markets in Brazil's cities. The burgeoning trade posed an additional threat to the Indigenous peoples and their traditional way of life.

Benki's fever subsided and that night he attended a pre-wedding ceremony for his son and future daughter-in-law. The event was held in a large traditional-style Puyanawa thatched longhouse, the interior of which was dirt-floored and lit up with many candles. There, with night falling, several hundred guests were greeted and invited to take ayahuasca. After this, silence reigned as everyone lapsed into their private mental experiences, which lasted several hours.

The next day, the wedding took place in a large round structure, followed by traditional dances by young Puyanawa and Ashaninka men and women. Benki posed for pictures with his son and new daughter-in-law, and the governor of Acre. He was a swaggering white man in loud branded clothing and was surrounded by half a dozen loutish-looking bodyguards.

Nature Worth Fighting For: Biopharmacy and Bioeconomy

One of Benki's followers grimaced at the governor's presence. He was a Bolsonarista, she confirmed, and was involved in a potentially disastrous agribusiness scheme to plant soya in Acre. Even so, she added, it seemed 'a good sign' that he felt the need to show up at the Indigenous event. I watched Benki's body language for any signs of discomfort at the governor's presence, but I saw none. One of his acolytes whispered to me loyally that Benki was 'all about alliances'.

After the wedding, Benki's schedule called for him to travel on to northeastern Brazil, and so he asked Federico Quitadomo, and Eliane Fernandes, his personal assistant, to accompany me to Yorenka Tasorentsi. We flew in a small prop-plane from the city of Cruzeiro do Sul, an old rubber-boom hub on the Juruá, to Marechal Thaumaturgo, a river traders' town further upstream, where Benki's reforestation project was located.

Yorenka Tasorentsi lay along a high bluff along the west bank of the Jurua opposite Marechal, its muddy banks lined with a floating honky-tonk line-up of general stores, bars and gas stations for the passing river traffic. Small passenger launches and motor-driven canoes made a racket as they came and went along the muddy river. Recent floods had torn away the trees that had lined the shoreline, and where piers had once bobbed in the water, overhung by shade trees, everything was an ugly raw mess of mud and precarious newly laid wood-plank walkways.

We reached Yorenka by riverboat, then carefully made our way up a slick mudbank. There, one of Benki's cousins, Leoneason Oliveira da Silva, or Leo, lived in a green-painted concrete house with his wife and children. Leo had worked for Benki since his youth, and he now ran Yorenka full time.

HOW TO SAVE THE AMAZON

Behind Leo's house stretched a managed jungle, much of it former cattle pasture, now reforested with native trees after various pieces of land had been successively bought up by Benki over the years. A dirt road led from Leo's house to several more structures, including a big tractor barn and a multi-storey house that belonged to Benki, its outer walls illustrated with ayahuasca-influenced murals of jungle creatures, Indigenous people and trees.

On either side of the road were several large fish-breeding ponds. Some of them were murky, however. With dismay, Federico explained that nearly two-thirds of the fish, intended to be the start of a commercial business venture, had been wiped out in the recent floods. He calculated the loss to Yorenka at close to a million dollars. At one of the ponds, workers were busy with a big yellow Caterpillar, moving earth around and building up a waterside berm in case of renewed flooding. Eliane explained that the Caterpillar had been purchased with US$50,000 Benki had been awarded from a French environmentalist society. In the near distance, seven sacred *sumauma* trees towered like ancient sentinels around Yorenka's headquarters.

Leo showed me the Yorenka greenhouse, where around 50,000 seedlings were being nurtured, everything from orange and açaí to mahogany and cedar. Next, he took me on an ATV tour through the lands being reclaimed; Eliane came along. In one especially scraggly area of scrub forest and pasture, Leo explained that it was a former cattle ranch that was one of Benki's most recent acquisitions. It had yet to be reforested and still had cattle roaming on it – they had gone wild. 'We're gradually eating them,' smiled Leo. 'The jaguars have helped too. And rustlers have also killed some.'

Nature Worth Fighting For: Biopharmacy and Bioeconomy

In another section of forest where the ground was choked with high grass, Leo explained that even after reforestation, cattle grass had persisted and they hadn't yet figured out how to get rid of it. Elsewhere, he showed where flooding had penetrated and then receded, leaving a layer of mud that hindered new plant growth.

In an area that looked parched, Leo explained that notwithstanding the problems caused by occasional flooding, drought conditions had been steadily increasing in the rainforest for years; he assumed it was because of climate change. 'There's less water in the streams than there was when I was a boy,' he said. He described coming across the carcasses of land tortoises and tapirs that had died of thirst. Leo was 38 and he had begun to wonder lately whether Acre's rivers would still have water when his children were his own age. 'It's all happening so fast,' he said, looking troubled. 'That's why Benki wants to plant a lot of trees, to *keep* the water, to correct the problems we humans have caused with our deforestation.' Even with the best of intentions, there were clearly numerous obstacles to rescuing the Amazonian wilderness.

Leo halted the ATV in a small clearing next to a raised wooden platform. Benki had selected the spot to build a guesthouse for foreigners who would pay to come to Yorenka for a 'detox diet and ayahuasca retreat', Eliane explained. She showed off the surrounding grove of planted trees and bushes that included guava, lemon, tamarind and ayahuasca. In its first phase, Eliane explained, Yorenka had focused on environmental restoration, but it now included 'spiritual healing' as one of its priorities.

The next day, Eliane and Leo took me downriver to a spot directly across from Marechal. It was where Benki had begun the Yorenka project 20 years earlier. Leo led the way through the forest there, pointing out trees that he had helped plant as a teenager. We came to a star-shaped stone-inlaid area which, he explained, had been purpose-built for ayahuasca ceremonies. There was also a wooden guesthouse and an amphitheatre for group talks and gatherings. In those early days, Leo said, a stream of VIPs had come to visit Yorenka, and in a ritual that had become customary, each of them would contribute by planting a tree. He patted the thick grey trunk of a tall tree that he said had grown from a sapling planted by Danielle Mitterrand.

Benki had chosen the site opposite Marechal to show the settler community there how the rainforest, which they were steadily destroying to carve out farms and cattle ranches, could be preserved by adopting a more sustainable lifestyle, like the Ashaninka were already doing on their forest reserve. 'At first, they were hostile,' recalled Leo. 'They said: "What's that indio doing *here*? He should go back upriver to his reservation."'

The Ashaninka reserve, where Benki's home village is located, was a four-hour journey by boat away, on the Amônia river, a tributary of the Juruá. Leo's family had been *seringueros*, rubber tappers, who had lived in the forest nearby, and his family had intermarried with Benki's. At 14, Leo had been sent by his family downriver to Marechal to attend school and when he was 17, Benki had invited him to join him at Yorenka.

Starting with Leo, Benki had persuaded more local youths to work on the reforestation project. They had earned room and board and little else; what united them was their enthusiastic faith in Benki, and over time, as the Yorenka project took root and important people came to visit, the ill-feeling in Marechal had seemed to subside. By 2009, 20,000 trees had been planted, recalled Leo, when a previous flood had surged along the Juruá and destroyed everything they had done. Afterwards, said Leo proudly, he and the other youths had carried on undeterred, and planted 150,000 more trees. In 2012, under Benki's direction, they had founded the Association of Young Warriors of the Forest. Leo was its president. The association still had a modest membership of 25, said Leo, but he held out hopes that it would grow further. 'Our hope is to have a big group that will work to protect the forest and the rivers,' he said.

I asked about the current relationship with Marechal. The two places seemed to represent very distinct social realities, and yet coexisted in close proximity to one another. By way of an answer, Leo said that Benki's team had offered to plant trees in the town but that the authorities had turned down the offer, saying that it would cause 'too much leaf rubbish'.

There were other, more serious problems, as it turned out. Benki's brother Isaac had moved on from his job as Marechal's mayor, and even though his successor was friendly to Yorenka, not all local officials were as open-minded. Eliane disclosed that a security officer had showed up uninvited at Benki's last birthday party at Yorenka, and threatened him with a gun. People had intervened and no shots were fired, but the episode seemed an ominous sign. Later, after the incident was reported to higher authorities, the officer had been transferred from Marechal, but he was still on active duty elsewhere in Acre.

Given everything, considering the man-made threats and those posed by climate change, I left Yorenka wondering whether Benki's vision of a reforested Amazonian utopia with fish farms and detox ayahuasca retreats was achievable. At the very least, it seemed an uphill battle.

Coincidentally, a few weeks after I visited Acre, the Harvard economist Ricardo Hausmann published an article with the provocative title, 'The Bioeconomy Will Not Save the Amazon'. In it, Hausmann took issue with what the idea 'that the best way to protect the Amazon is to cultivate a "bioeconomy"' that will 'foster the sustainable use of forest resources and promote the welfare of local communities . . . [and] counter destructive practices that contribute to deforestation, such as cattle ranching. Although well-intentioned, this approach is likely to backfire.' Pointing out that the main crops of a potential Amazon bioeconomy, such as açaí berries and Brazil nuts, 'represent niche markets valued at about $1 billion, or roughly 0.05 per cent of Brazil's GDP,' Hausmann wrote, 'such a small market cannot sustain the Brazilian Amazon's 30 million inhabitants.' Moreover, he concluded, any efforts to create a bioeconomy at scale would very likely accelerate the destruction of the Amazon by 'attract[ing] resources and people into the rainforest instead of driving them away . . . A better way to protect the Amazon would be to boost the productivity of the region's urban centres and surrounding non-forested areas. Given that most people prefer the comforts of urban living to the hardships of forest life, this strategy moves individuals seeking stable and quality jobs from forested regions to the cities.'

Even if Hausmann's arguments against the bioeconomy were accurate, his vision of a rural exodus into the cities sounded unrealistic. In his essay, Hausmann prominently cited Marek Hanusch, a German economist for the World Bank. In 2023, after several years conducting research in Brazil's Amazon, Hanusch had published an extensive study entitled 'A Balancing Act for Brazil's Amazonian States: An Economic Memorandum'.

Nature Worth Fighting For: Biopharmacy and Bioeconomy

Hanusch had been explicit about the stark realities of the current Amazonian crisis. 'In the shorter term, there is an urgent need to halt deforestation – a massive destruction of natural wealth that poses risks to the climate and economy,' he wrote. 'Amazônia is Brazil's deforestation hot spot, and the Amazon rainforest is approaching tipping points into broad and permanent forest loss. Reversing the recent increase in deforestation requires stronger land and forest governance, including land regularization and more effective law enforcement.'

To save the Amazon, Hanusch argued, 'a new growth model' was needed for Brazil and the Amazon, 'a structural transformation' that would gear its economies away from resource extraction and towards productivity by revitalizing its urban economy based on manufacturing and services. This could attract more people to towns and cities and away from 'the agricultural frontier', he wrote. 'The public-good value of Amazônia's forests could generate conservation finance linked to verifiable reductions in deforestation. Such financing would support a new development approach, combining forest protection, productivity, balanced structural transformation, sustainable production techniques (including the bioeconomy), and other measures to address the needs of Amazônia's urban and rural populations. This approach must also heed the needs and interests of Amazônia's traditional communities.'

Still unsure that I understood what his arguments meant for an Amazonian bioeconomy, I telephoned Hanusch, who is now the World Bank's lead economist in East Africa. I told him of my trip to Acre and what I had seen of Benki's efforts at creating a bioeconomic start-up model at Yorenka. Hanusch knew of Benki and thought highly of him, he said, but believed bioeconomy projects like his could only work on a limited scale. 'Bioeconomy is a utopia that defies the logic of economics,' he said. 'A red herring. The idea that this is the kind of thing that will save the Amazon is just nonsense. Unless you can buy *all* the Amazon, it will not work. If you remove productive land to reforest it, then it will just be deforested elsewhere. Also, if you monetise land that's been deforested, you create value for the land, and it's an incentive for more land-grabbing. The only way you can stop deforestation is by changing your economic model. Right now, the only kind of viable economy in the Amazon is agribusiness, and if you want economies of scale, you are going to have to change the nature of the forest, and that's what's already happening. At the end of the day, deforestation is a macroeconomic choice, and so long as Brazil's growth model is based on agriculture, you're going to see expansion into the Amazon. But if you build up the infrastructure in the cities, you'll get people moving into them.'

By way of an example, Hanusch recalled how, early on in his Brazil research, he'd been dumbfounded to find a motorcycle assembly plant in the remote Amazonian city of Manaus. 'I used to think: Why are they producing motorcycles in a place like Manaus? It seemed crazy to me. But now I've come to think it's a good thing because they're not using the forest to make them.' More of that was what was needed.

In the end, Hanusch's arguments for a viable Amazon depend on an evolution in government policy circles, which, in turn, depend on having a government in place that cares about conservation and believes in climate change, and institutionalises such policies over the long term. That might be possible in some countries, I thought, but was it possible in a volatile democracy like Brazil? Following the calamitous four years of Bolsonaro, Lula's return to the presidency in 2023 offered a restoration of hope to environmentalists that new, forward-looking policies could be implemented on behalf of the Amazon. Lula is no tree-hugging environmentalist, but at least he would not be trying to actively burn down the Amazon rainforest. A promising sign was his call to Marina Silva to become his environment minister once again. She had served in the same role during his first term in office but had resigned in disagreement with his desire to balance conservation with development. Now, however, Lula had promised a zero-tolerance policy on deforestation; in an interview we had soon after his election victory, he reiterated that goal, but also spoke, somewhat hedgingly, about the need to plan for 'sustainable development' on behalf of the thirty million Brazilians thought to be living in the Amazon. In Brasília, a few weeks before Lula assumed the presidency again, Silva told me that even with the new government in place, getting things right in the Amazon would take time. 'We won't be able to do this in four years,' she said. 'The problem during Bolsonaro was that the transgressors had total impunity. With Lula, at least, the *expectation* of impunity will end.'

In Brasília, I met with Carina Pimenta, Brazil's national secretary for the bioeconomy. A personable woman in her forties, Pimenta smiled sympathetically when I recounted my trip to Acre and shared my ongoing confusion over the definitions of bioeconomy. She agreed that the topic eluded easy definition, but she was clear what it meant to her: using Brazil's biodiversity to create sustainable economies that would not hurt its environment. Pimenta seemed well suited to her job, having managed an investment fund for sustainable Amazonian development for several years before starting her own NGO to work with grassroots communities.

Pimenta acknowledged that there were numerous challenges to creating an Amazonian bioeconomy of scale based on Brazil's ever-expanding capitalist growth and development model. 'This will take a while,' she said. 'We need to ask ourselves where we will be thirty years from now. It is important that different models emerge.' Pimenta said she had drawn inspiration from the Indian economist Amartya Sen. 'Sometimes small economies are good because they can be successful if they are well targeted. Maybe we can develop small biodiverse-friendly economies and then look at the bigger solutions.'

Among other schemes, Pimenta was lobbying for Brazil's government to build an information bank on Amazonian biodiversity. The Brazilian state had never invested in such a project, she pointed out, but she believed it was essential to begin one. She was sceptical that carbon-offset schemes represent long-term productive components of a genuine bioeconomy, asking, 'Is storing carbon an economy? What about *knowledge*?' Indigenous communities had a great deal of stored-up knowledge about medicinal plants and natural remedies to share, she said; this knowledge would be invaluable in the development of an Amazonian biopharmacological economy. As an example, she mentioned drug trials taking place in the Amazon with a natural antidepressant that had been discovered by an Indigenous community. Through benefit-sharing arrangements between Brazil's government, pharmaceutical companies and Indigenous communities, future positive results of such trials could help lay the groundwork for an Amazonian bioeconomy that would benefit Brazil commercially as well as help preserve its rainforests.

Pimenta enthused about another programme Marina Silva had fostered, involving government support for environmentally friendly businesses in a series of 18 'biodiversity hubs' that had been designated across Brazil. And there were other promising developments. Pimenta had attended the 2023 G20 summit in New Delhi where, she pointed out with a note of pride, the term 'bioeconomy' had been introduced for the first time. And at the Amazon Summit held in Belém that same year, convened by Lula for the leaders of eight Amazonian nations, Pimenta recalled with a rueful smile: 'There was a lot of dispute about what exactly the bioeconomy *is*, but the one thing everyone agreed on was that it is worth fighting for.'

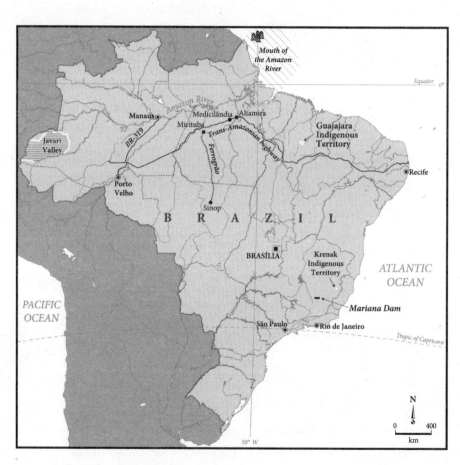

Selected mega-projects, completed and under consideration

CHAPTER TEN

A LIFE-CHANGING RELATIONSHIP: EDUCATE AND RETHINK

Jonathan Watts and Dom Phillips
Javari Valley, Amazonas

Jonathan Watts is a journalist who lives in the Amazon rainforest municipality of Altamira. He is The Guardian's *global environment writer, founder of the trilingual rainforest-based newsletter Sumaúma.com, initiator of the Rainforest Journalism Fund, and author of* The Many Lives of James Lovelock *and* When a Billion Chinese Jump.

'The best teachers are the Amazon's original inhabitants: its Indigenous peoples.'

When all is said and done, after Dom's life-changing, life-ending search for Amazon solutions, after his years of scrabbling for travel funds and publishing deals, after epic river expeditions down the Solimões, the Tapajos, the Negro, the Xingu and Tefé, after dusty drives through Pará, Acre, Amazonas, Roraima, Maranhão, Rondônia and Mato Grosso, after leave-no-stone-unturned efforts to track down the people who might have answers, after the reading of thousands of pages of reference books, magazines and news websites, after hundreds of beers and coffees and chinwags, and after long hours in front of a computer screen trying to condense it all down into crackling, vivid prose, Dom left us with a big unanswered question: what was his conclusion?

Dom's final chapter was tentatively titled 'Educate and Rethink'. His notes suggest this was the most amorphous, ambitious and still undeveloped of all the sections. He had not drafted a word, although his outline mentions fake news, propaganda, religion, philosophy and education. Exactly how he expected to bring these strands together is not clear. Perhaps he had not yet solved this riddle himself. He left clues in the form of half-suggestions: 'Finishes on the scene in Medicilândia?', 'Close with the day with the students and professor.' But what happened in Medicilândia, and who were these people? Many of his notes, often written on speedboats or bumpy roads like the Trans-Amazonian highway, were practically unintelligible, even after multiple attempts by family and friends who know his handwriting best.

To grapple with this conundrum – and the profound but abstract ideas at its heart – I wrote a note of my own: 'relationships'. Dom, it seemed, was considering different ways to make readers both outside and inside Brazil feel more connected to the Amazon and its people, and by doing so, to change their thinking about its value. As my principal guide, I focused on two aspects of his approach: first, the strengths and limits of journalism, and second, his sensitivity and receptiveness to new ideas, encapsulated by another heading in his outline: 'Listen to Indigenous People'. The world, he wrote, 'is not a disconnected, random series of nations and societies, but an interconnected whole whose survival depends on cooperation, not competition.' To understand this, he argued, 'the best teachers are the Amazon's original inhabitants: its Indigenous peoples.'

A Life-Changing Relationship: Educate and Rethink

But how had he come to this understanding, which was such a long way from his suburban roots in the Wirral and his early career as a musical journalist? And how was he planning to talk about it in his final chapter? I would never know what he planned to say, but I could try to trace the intellectual, as well as physical, journey that he had made. What were the ideas that had drawn him to Bruno, and taken them both so far away from their homes to the Javari Valley, with all its risks? What connections were they looking for? What did they find and follow? I decided to work my way backwards, starting with the last place they had visited.

Considering its location on the frontline of a deadly conflict, the Lago do Jaburu feels a blessedly tranquil place. A couple of archetypal riverine homes, built on stilts, with wood-plank walls and corrugated roofs, are perched at the top of a steep bank above the Itaquaí river. On one side of these buildings is a grassy depression that turns into a channel during the rainy season. To the rear is the serene lake, surrounded by thick vegetation, that gives this community its name. Brightly coloured wooden canoes can usually be found tied to the branches of the nearest tree. At mid-morning, with the sun high in the sky and temperatures already creeping above 30C, the only sounds are birds and the occasional splosh of a fish flopping momentarily out of the water.

This unlikely idyll, which is less than a kilometre from the demarcation line of the Javari Valley Indigenous Territory, was where Dom and Bruno spent their last day. They had come here to join the Indigenous surveillance team that patrolled the border between the protected territory and its hinterlands. Having traced Dom's final journey, I could imagine how relieved he would have been to reach this spot. After a tense river journey past several hostile communities of fishers, here he could feel relatively safe. He and Bruno were meeting friends. And, given his search for Amazon solutions, he would surely have been cheered by the positive change that had come over the Lago do Jaburu.

The lake had once been so over-fished that it was considered almost lifeless, but a resource management plan had been put in place more than a decade earlier and maintained by the old man who lived there, known by everyone simply as Raimundinho. The *pirarucu* were now flourishing, thanks to his efforts. And his courage – Raimundinho had received death threats from local fishers who wanted quick money from catching and selling the fish. But the white-haired *ribeirinho* had stood his ground. He had grown tired of running down nature and the endless conflict it brought with his Indigenous neighbours. The best way to avoid that was to stay out of their territory and, instead, to carefully nurture the fish stocks on the lake by his home. Now the lagoon was once again full of life.

In the morning and afternoon of that last day, Dom – as rigorous as ever – individually interviewed all 13 men on the surveillance team, asking them the same questions: how did they protect their territory, for whom were they protecting nature, in what way were they affected by the political situation? After dark, he and Bruno joined a simulation of a monitoring mission, crossing the river on small boats and forging quietly through the forest amid a chorus of insects and frogs. Dinner that night was a feast of sloth and *pirarucu*. The next morning, they shared a breakfast of grilled fish and black coffee, then Dom and Bruno thanked and hugged everyone individually, before setting off on that last fateful journey.

A Life-Changing Relationship: Educate and Rethink

A member of the team, Higson Dias Kanamari, of the Kanamari people, recalled his last encounter with Dom with a mix of affection and horror. 'He was very happy to be among us Indigenous people,' he said. 'When he was with us, he had a second family. We looked after him. I could see the pleasure he had from being with us. Unfortunately, we couldn't anticipate the extent of the evil that people wanted to do.'

Higson talked to me about what happened after Dom and Bruno left the lake. For weeks after their deaths, the nearest town of Atalaia do Norte was flooded with reporters. Residents wryly noted that the intense coverage of the death of a white foreign journalist was in striking contrast with the murder three years earlier of FUNAI officer Maxciel Pereira dos Santos, who had worked closely with Bruno in tracing illegal fishing and hunting operations. In September 2019 Maxciel was shot dead. His family believes the assassination was carried out by the same people who killed Bruno and Dom. But nobody was ever charged and the case barely made a ripple outside the region.

As well as the double standards, Higson said the treatment of the two crimes showed the power of stories that can attract a global audience. 'When they killed Maxciel, nothing happened. But with Dom and Bruno, there was an enormous interest.' He saw this as a positive: 'The media was the focal point for the world to learn about the defenders of the forest. That forced the government to respond. The Brazil brand required the state to show it had a responsibility to care for its land and its people.' For Higson, this created a powerful connection not just with the government in Brasília but with people in other countries. It was a way of changing people's minds, of building relationships, of closing the gap between the forest and the city. 'We lost two important people but we gained allies in the world. That is oxygen for us,' he said.

I guess Dom would have felt more than a little embarrassed by this. He would also have been conscious of a certain unfairness in the treatment of people, even after death. Like the US nun Dorothy Stang, Dom was an unusual Amazon martyr in being white and from a rich nation. Most of the others were Indigenous, *quilombolas*, *ribeirinhos* – victims of murders that were never investigated or covered by the media, people whose names and faces were largely unknown outside their home towns. It is a similar story across the world, where more than 1,900 people have been murdered since 2012 trying to protect their land and resources. That's an average of one killing every two days.

Dom and Bruno became hyper-visible as symbols of that undeclared war on nature.

When I got home from Javari, I realised I needed to know even more about Dom's later thinking about the Amazon, and how that aligned with Bruno's. My own friendship and working relationship with Dom gave me some insights, but I needed to talk to other people.

My first meeting with Dom was in mid-2012, sitting outside a sidewalk *boteco* in Jardim Botânico on one of those balmy Rio de Janeiro nights that are made for sipping caipirinhas or ice-cold beers. He had reached out because we were both British journalists who had recently moved to the city. Dom, who already spoke excellent Portuguese after several years in São Paulo, was settling in much more quickly. I was the recently appointed correspondent for *The Guardian*, still struggling with the language of my new home and finding everything in Brazil absurdly bureaucratic, slow-moving and expensive. Dom, on the other hand, appeared to have already got under the skin of the country. He was infectiously enthusiastic about its music, sport and culture, and seemed fascinated by the politics.

What I recall most clearly was his exposition on *alegria*, a sense of joyfulness, that he felt was a characteristic of his adopted country. He told me too about his book on the DJ club scene in the UK in the 1990s, and, by contrast, how chemically fuelled happiness and idealism had soured as money and drugs took over. Dom said he was happier now in Brazil, even though he was sometimes struggling to make ends meet as a freelance journalist stringing for a bevy of publications on everything from politics and football to fashion and the oil industry.

Over the years, I came to feel he was always looking for a deeper understanding. Maybe this was natural curiosity. His interest in the world was like a mental searchlight forever scanning the horizon. Whether in a press conference or a bar, if he thought someone had anything interesting to say, he would fix them with his piercing blue eyes and begin a gentle but relentless interrogation. In the years that followed, we connected through a shared love of Bowie, Björk and Fela Kuti, and a passion for nature and outdoor sports.

Scrolling back through WhatsApp archives, I realised many of the stories and pictures Dom shared were of spectacular views or wildlife he encountered: rays, whales, turtles and sharks seen during standup paddleboard outings around the coastline of Copacabana; capuchins, marmosets and toucans encountered on hillside walks around Rio de Janeiro. With a group of like-minded friends, we made weekend hikes through the mountains between Teresópolis and Petrópolis, climbed the Pedra da Gávea to enjoy its stunning view of Rio and trekked up the slopes of Itatiaia for its breathtaking panoramas. More frequent were the bike rides. Early weekday mornings, a group of us would start the day with a cycle up to the Cristo Redentor statue on the Corcovado, a lung-busting activity that became known as 'Christ on a bike'. Dom's other great passion was culture. He was a regular at the São Paulo Biennial, Paraty literary festival and Rio music concerts. Back then, the Amazon was a remote concern. 'In those early years, he didn't talk much at all about the environment or Indigenous people,' his widow, Alê, said.

In September 2015, Dom made a reporting trip to the Amazon region for an article for the *Washington Post* on a team of forest guardians set up by the Guajajara, an Indigenous group in the western Amazon state of Maranhão. It was an eye-opening trip to one of the most scarred landscapes in Brazil. Maranhão has been treated as a fiefdom for generations by the Sarney family, an oligarchical dynasty of 'rural colonels', media owners and political heavyweights, including a former president of Brazil. Under their control, this state was substantially stripped of forest cover by an influx of loggers, ranchers and land grabbers. The Guajajara and other Indigenous groups had been steadily pushed into smaller and smaller enclaves, which were also the last vestiges of primary forest. Even these were now being threatened, so the Guajajara had organised themselves into monitoring and protection teams, which patrolled their territory and chased off invaders, sometimes even tying them up and carrying them out, after burning their chainsaws and trucks.

This was risky because the loggers were often better armed and had more political influence in local government. One of the

A Life-Changing Relationship: Educate and Rethink

guardians that Dom encountered, Paulino Guajajara, would later be assassinated in an ambush by armed loggers in 2019. The threats and violence have continued ever since. This brave defence of the land was emulated by other First Peoples and earned widespread respect. In 2023, Sonia Guajajara was nominated as Brazil's first Indigenous minister.

Alê said her husband was inspired: 'When he came back, he was absolutely in love. He was very impressed with the Guajajara's style of life and their courage in defending the Amazon. It really surprised him,' she recalled. 'I think this was the moment when Dom started to understand that land is far more than a piece of property with a market value.' From that moment on, he started to read more about the Amazon and make connections with Indigenous groups, land rights organisations and non-profit groups like ISA (Instituto Socioambiental) and Survival International.

His environmental awakening took on an extra dimension two months later when he was one of the first foreign correspondents to witness the effects of the most devastating mining disaster in Brazil's history. The collapse of a tailings dam in Mariana operated by Samarco, an iron ore joint venture between two of the world's leading mining conglomerates, Vale and BHP Billiton, had unleashed a torrent of toxic sludge, killing 19 people, leaving hundreds homeless and ruining the livelihoods of the Krenak Indigenous community. Never before had Dom seen an industrial calamity on such a scale. In a dispatch for *The Guardian*, he described 'apocalyptic images' and harrowing tales of lives turned upside down by 40 million cubic metres of contaminated liquid that 'polluted the water supply for hundreds of thousands of people, decimated wildlife and spewed a rust-red plume of mud down the Doce river'.

Dom had visited the area just a couple of months earlier on a holiday with Alê. Now he was seeing the streets they had happily sauntered along wiped away, the village beside their hotel destroyed and bodies lined up on the roadside. 'The dam collapse had a huge impact on Dom,' Alê said. 'He told me it was so sad, so terrible and so unfair because the company had known there were problems. From that moment on, he considered (the mining company) Vale a monster. We talked about it right up to his death.'

While such experiences were changing Dom's thinking in profound ways, his growing interest in environmental issues was not so much a Damascene conversion as a gradual change shaped by both pragmatism and idealism. After the Rio Olympics closing ceremony and the impeachment of President Dilma Rousseff, both in 2016, his freelance commissions from the *Washington Post* started to dry up. Editors there felt the story had moved elsewhere. To fill the gap, he took on more work from *The Guardian*, where a temporary vacancy had opened up after I left Rio to return to head office in London in 2017. Editors there, including me, were interested in the environment story. Dom, who had already been moving in that direction, saw an opportunity. He sent in the pitches thick and fast. And he ventured further than ever.

And after 2018, he had a guide. As we have read in these pages, that was the year Dom met Bruno and joined him on an expedition deep inside the Javari Valley to track the movements of an isolated group. The 1,020km trek was gruelling even for Dom, who was an indomitable hiker. Bruno made a huge impression on him. Dom sent a WhatsApp message to longtime friend

A Life-Changing Relationship: Educate and Rethink

and filmmaker Otavio Cury telling him the experience had been 'profound'. Two years later, he would tell him it was 'a trip that is still resonating with me today, an experience I will never forget. For that, I'm deeply grateful, though at the time I was largely wet, dirty, tired and quite often scared as they killed another snake like we swat mosquitos.'

Dom's sister Sian remembers him showing photos from that trip soon after. 'That was the first time I heard him talk about the Amazon and Indigenous people. Don't forget, he had previously been working for an oil industry publication. But that trip with Bruno really moved him. It was like he had been transformed.'

The 'Mr Cool' of the club scene had found someone far cooler doing work of profound importance with Indigenous activists in the vanguard of a globally important struggle. Alê believes this is when the idea of an Amazon book started to form in his mind. 'It's another world, Alê. You have no idea,' he told her, with awe.

After that trip, he and Bruno would often talk on the phone, exchanging contacts and ideas. 'All he could talk about was Bruno,' Alê said. 'He would say things like, "Alê, he has an Indigenous soul. He talks their language. He sings their songs and walks in the forest like them. He is afraid of nothing. You must meet him. I've never met anyone so committed."'

The relationship had been formed. Dom the disciple. Bruno the mentor. But how had the mentor himself come to care so much for the forest and its people?

Bruno was an urban child, born in Recife, the capital of the northeastern state of Pernambuco, 4,000km from the Javari Valley. Like Dom, there was little in his family background to suggest he would one day devote his life to saving the Amazon. His father, Max, was a sales executive for an aluminium and glass manufacturer. His mother worked in a government pension office. But family members said the young Bruno expressed an early connection with nature. Large parts of his childhood were spent at the Boa Viagem beach in Recife and his holidays were taken in the countryside at his maternal grandparents' house in Pilar, Paraíba state. His mother said he dreamed of working in the interior.

After graduating from the local college, he entered a journalism course at the Federal University of Pernambuco in 2000 – his interest in the influence of the media would later draw him to Dom and other foreign correspondents. But, like Dom, he dropped out of university, tried a number of different jobs, including at the social security office in Recife and several years at the Balbina hydroelectric power plant on the Uatumã river, in the northeast part of the state of Amazonas, where he coordinated reforestation activities and learned how to walk in the forest with the woodsmen.

This became a vocation. He joined the government's National Foundation of Indigenous Peoples (FUNAI). In 2010, he was assigned to the regional coordination office of the Javari Valley. He was thrilled by its reputation as a remote, challenging outpost where Indigenous groups still controlled vast tracts of land. When he arrived, he boarded canoes that took him upriver to the villages, contradicting the wishes of his then boss, who wanted him to stay at the head office in Atalaia do Norte. The Ituí River was one of his first destinations, visiting the villages of the Marubo people. Then came a visit to the Kanamari, who invited him to participate in a ritual using ayahuasca.

A Life-Changing Relationship: Educate and Rethink

He learned to communicate in five Indigenous languages: Kanamari, Marubo, Matsés, Matis and Korubo, and, as Dom noted, could sing community songs. Video clips show him crooning '*Wahananarai, Wahananarai*' in a sweet tenor with a smile on his face. That Kanamari song about a macaw and its young would become a lullaby for his own children, to whom he gave Indigenous middle names. Bruno chose his close Indigenous friend Beto Marubo to be godfather to his sons Pedro Uaqui (a Matsés name) and Luiz Vissá (a Korubo name).

'He was family to those people as those people were family to him,' said a close friend, the journalist and Indigenous rights campaigner Helena Palmquist. 'Bruno was big in size, big in laughter, big in courage. He wouldn't take no for an answer. That is why he could do such important work in FUNAI. He could be stubborn but he was always a sweet guy.'

Soon after he arrived in the region, he briefly met his future partner, the anthropologist Beatriz de Almeida Matos. They would reconnect a few years later, while Beatriz was completing her doctoral research and Bruno, then coordinator of the regional FUNAI office, was fending off an invasion of his office by supporters of an Indigenous leader who he had exposed for selling off hundreds of turtles. 'It was at that time of tension and total confusion that we became close,' she recalled. 'We spent days and days talking together. We were both visiting Indigenous territories, really living with them and learning their ways. I think talking to me helped calm his mind in the middle of the conflict. Somehow, it worked. I moved in with him.'

In the years that followed, they would travel often into the Javari Valley and sometimes live in the forest communities for extended periods. Their knowledge of Indigenous culture deepened, and with this understanding came increased respect for Indigenous ways. 'When you engage profoundly, you go on hunting expeditions with the women and their partners, then you can see how their relationship with animals and the forest is different. It is about co-existence, about preservation, about equality,' Beatriz explained. 'It is very different from the European vision of dominating nature. It is very profound and much more spiritual. When they hunt animals, they care for their spirits. I came to admire them deeply.'

In 2012, Bruno took over as FUNAI's regional coordinator of the Javari Valley. Over the following four years, he established a new surveillance base on the Curuçá River and encouraged Indigenous people to vote and participate in politics. In 2012, just one Indigenous person was elected to the city council. Four years later, there were five, out of a total of eleven.

Bruno's priority was the protection of uncontacted and recently contacted peoples, who were threatened by drug traffickers, illegal loggers and poachers. This became a passion, according to Palmquist: 'When we got together over a beer, we never talked about how to protect the forest. It was all about preventing genocide.' This important distinction was often lost on the outside world. Foreign environmentalists tend to focus on Indigenous people as a means to an end – the conservation of the Amazon biome as a globally important carbon sink and climate stabiliser. But for those like Bruno who worked in the villages, it was the other way around. 'We save the forest because it is the place where Indigenous people live,' Palmquist said.

Beatriz said this shaped his vision of how to preserve the Amazon and live with nature – a belief that he wanted to spread as widely as possible. This later goal was informed by his interrupted journalism training. 'He knew the importance of telling stories to get funds to protect isolated people,' Beatriz recalled. Foreign journalists seemed most receptive.

'Bruno knew how to work with them, how to communicate,' Palmquist said. 'We talked about that. We thought the domestic media is harder to work with because of political and money commitments. So, we have to rely on foreign media, even though that also poses problems because they like to romanticise. We used to joke that "they always go to the heart of the Amazon. Nobody ever goes to the liver".' Bruno made great allies. This is where Dom came in. That first expedition in 2018 was the start of a relationship. 'They were friends. Without a doubt,' Beatriz said. 'They trusted each other.'

Bruno's embrace of the Javari cultures and his desire to work on the ground rather than at the desk earned him the respect of the older generation of Indigenous experts, such as Sydney Possuelo, the former FUNAI chief. Possuelo is among a small band of men – known as *indigenistas* – who have an almost mythical status in Brazil due to their courage in venturing deep into the forest to engage with Indigenous communities and their strong moral stand in promoting recognition of their territories and a hands-off approach to peoples who choose to remain in relative isolation. Around 2015, Possuelo started to invite Bruno on expeditions, which was seen by some as anointing him as a successor. Three years later, Bruno was put in charge of the central government's isolated peoples programme, effectively making him the leader of a new generation of indigenistas.

As this book has detailed, everything changed in 2019 when Bolsonaro took power. FUNAI was dismantled from within.

A former police chief, Marcelo Xavier, was appointed as its head and, the following year, a fundamentalist evangelical, Ricardo Lopes Dias, was made chief of the Isolated People's Department, with a primary goal of opening areas like Javari to missionaries.

Bruno was sidelined. He took a period of extended leave and collaborated with Orlando Possuelo (Sydney's son, who was then based in Atalaia do Norte), Beto Marubo and others to set up a new group called the Observatory for the Human Rights of Isolated and Recently Contacted Indigenous People (OPI) with representatives of the Indigenous peoples in the Javari Valley. The group, formally established in 2022, effectively became a shadow FUNAI, filling in for a government that had gone AWOL. It organised the *equipe de vigilância* (monitoring team) to fill the protective role the state had abandoned. As well as patrols, it ran training programmes and used GPS trackers, drones and camera traps to gather evidence of intruders who illegally entered the protected area to fish, hunt, mine and explore the land for ores and minerals.

Nationwide, the situation was growing grimmer. In Bolsonaro's first full year in power, Amazon deforestation hit the highest level in more than a decade. Thousands more *garimpeiros* (illegal miners) flooded into Indigenous territories. Invasions of protected areas became more commonplace. In a message to me in 2019, Dom expressed deep dismay about his adopted homeland. 'In 12 years here, this is the low point. I have never felt so bleak about this country's future,' he said.

Dom reported on the many ways Bolsonaro used lies to influence public opinion. To make fake news stick, the president had to cull truth tellers. He fired the director of the country's prestigious

A Life-Changing Relationship: Educate and Rethink

National Institute for Space Research, Ricardo Galvão, after the institute published a report that concluded that the number of forest fires in Brazil had increased dramatically. He then, outrageously, blamed environmental groups for arson attacks. There was not a shred of evidence for this allegation, but the bold lie pleased his base and distracted the mainstream media. He followed this up with ever more outlandish untruths. 'This story that the Amazon is on fire is a lie,' Bolsonaro said during the 10 August second Presidential Summit of the Leticia Pact for the Preservation of the Amazon. 'And we must combat this with real numbers.'

Of course, the real numbers continued to show devastation. During Bolsonaro's four years in power, there was a 60 per cent increase in deforestation, the largest increase ever recorded in a presidential term. Brazil's green greenhouse gas emissions rose 12.2 per cent from 2020 to 2021, the highest in 19 years. Meanwhile, there was a 38 per cent decrease in the number of fines for environmental crimes. The far-right government was letting this happen, yet it not only denied responsibility, it noisily blamed others.

In a plan for this final chapter, Dom outlined his intention to confront this:

> His [Bolsonaro's] propaganda machine, coordinated by young, social media savvy, far-right militants in a room inside presidential headquarters called the 'Hate Cabinet' by Brazilian media, continues to bombard Brazilian social media networks. The propaganda works. Plenty of Brazilians suspect NGOs started fires to get foreign money, simply because the president said

so without ever providing any evidence. One old lie widely propagated during Brazil's military dictatorship [1964–1985], shared by Bolsonaro and still voiced by military officers today is that foreign powers seek to help Indigenous people set up independent states on rainforest reserves to access Amazon riches.

Dom's views on Bolsonaro were shaped by personal experience. In July 2019, he asked the president about the surge in forest fires and drew a fierce response. He remembered the exchange as follows:

> Covering for *The Guardian*, I attended a televised press conference for international media and was positioned two chairs from Bolsonaro. When I asked how, given rising Amazon deforestation figures, he could convince the world that Brazil was serious about protecting the Amazon, Bolsonaro launched into a belligerent, rambling rant. 'First you have to understand the Amazon is Brazil's, not yours,' he said, questioning his own government's deforestation data and suggesting that the technician responsible was being paid by a foreign non-profit group. 'We preserve more than everyone; no country in the world has the moral authority to talk about the Amazon.' Within an hour, the clip of this interchange was released on a pro-government site, headlined: 'Bolsonaro detonates foreign journalist.' It went viral, swirled around WhatsApp groups, was shared by Bolsonarista politicians – viewed 446,00 times on congresswoman Joice Hasselmann's Instagram.

Bolsonaristas frothed in approval. 'Tell this scum to take it up the you-know-where,' one supporter posted. 'Congratulations president.'

Dom's question was also published on Bolsonaro's official profile under the title 'The false defence of the Amazon by other countries'. Dom had been targeted by the most powerful man in Brazil. He told me that he felt he had been set up and that the president was making life more dangerous for journalists in a bid to win the propaganda war. It also directed him to another key message for the last chapter of his book: 'A fundamental shift in protecting the Amazon involves convincing people how this is in their own interest – and that is a challenge in Brazil where, for many, environmental protection is a vague concept easily distorted by fake news and appeals to nationalism.'

When international criticism intensified, Bolsonaro and his supporters responded with fresh false allegations and outrageous statements guaranteed to steal headlines. After Pope Francis and others invoked the 'blind and destructive mentality' of those behind the devastation of the rainforest, Bolsonaro claimed, 'Brazil is the virgin that every foreign pervert wants to get their hands on,' referring to the Amazon. When French President Emmanuel Macron expressed outrage about what was happening in the Amazon, pro-Bolsonaro social media groups spread unsubstantiated rumours about a secret alliance between Macron and Brazilian politicians: 'We have traitors in our midst; they are in a secret alliance with foreign leaders to give the Amazon to foreign powers.' In a 2019 speech to the UN General Assembly after one of the worst years of Amazon fire in recent history, Bolsonaro insisted the forests were 'practically untouched' and blamed a 'lying and sensationalist media' for exaggerating their destruction. In any case, he told the world, it was none of the business of outsiders because the Amazon is a Brazilian resource rather than 'a heritage of humankind'. All of this was aimed at creating distance. The last thing the Brazilian president wanted was an emotional connection between the outside world and the rainforest because that might interfere with exploitation schemes.

Bolsonaro also used the UN General Assembly to attack the most influential Indigenous leader in the Amazon, Chief Raoni Metuktire. He declared him a puppet of foreign powers.

Following Dom's approach to 'listen to Indigenous voices', I sought the opinion of Raoni on Dom's thinking. The elderly Kayapó man said Dom and Bruno were heroes: 'They were dedicated to the struggle of Indigenous people and efforts to monitor and protect our territory. They put all their efforts into the forest, despite the threats from drug traffickers and illegal land grabbers. The Kayapó, and all Indigenous people in Brazil, were very sad when we heard of their deaths.'

Asked if anything positive could come out of such a horrendous crime, he answered: 'They gave global visibility to the Indigenous struggle. The world could see what we have to face. We will continue the work they were doing. And other big-hearted, non-Indigenous people will join us. It is a cause that anyone can embrace.'

Raoni has been campaigning for the forest for almost all of his 90-plus years. At first, he fought with spears and bows, but he came to realise the battle for ideas was far more important. Thanks to his lip disc, beads, earrings, flowing grey hair, trenchant statements and friendship with global celebrities like the musician Sting, he is probably the best-known Amazonian in the world. Raoni was a young, jenipapo-painted warrior when his people were contacted in the early 1950s by non-Indigenous invaders, who brought gifts of metal blades and beads but left behind European diseases such as malaria, influenza and measles that decimated the population. In the 1970s and 1980s, Raoni was among the leaders of the often deadly fight against the BR-080 road, cattle ranchers and the Belo Monte dam. He was fêted by world leaders and met the pope, gaining a level of prestige and leverage that challenged the prejudices of the many Brazilians who saw Indigenous people as poor and uneducated. This helped the Kayapó to secure government recognition of their territorial rights across a vast chain of protected lands, which formed the spine of a north–south firewall against deforestation.

After Bolsonaro took power, Raoni felt it was time to go back to this war of words and feelings. 'From many years ago, I fought in campaigns and appeared in the media. Then, when we won the victory of having our lands demarcated, I stopped because everything seemed fine, everything was tranquil,' he told me at the time. 'But the president threatens Indigenous people, so I came back to fight again.'

He urged the outside world to change course or face dire consequences:

> We call on you to stop what you are doing, to stop the destruction, to stop your attack on the spirits of the Earth. When you cut down the trees, you assault the spirits of our ancestors. When you dig for minerals, you impale the heart of the Earth. And when you pour poisons on the

land and into the rivers – chemicals from agriculture and mercury from gold mines – you weaken the spirits, the plants, the animals and the land itself. When you weaken the land like that, it starts to die. If the land dies, if our Earth dies, then none of us will be able to live, and we too will all die. So why do you do this? We can see that it is so that some of you can get a great deal of money. In the Kayapó language we call your money *piu caprim*, 'sad leaves', because it is a dead and useless thing, and it brings only harm and sadness . . . But those rich people will die, as we all will die . . . You have to change the way you live because you are lost, you have lost your way. Where you are going is only the way of destruction and of death. To live, you must respect the world, the trees, the plants, the animals, the rivers and even the very earth itself. Because all of these things have spirits, all of these things are spirits, and without the spirits the Earth will die, the rain will stop and the food plants will wither and die too. We need to protect the Earth. If we don't, the big winds will come and destroy the forest.

Then you will feel the fear that we feel.

Indigenous intellectuals have published some of Brazil's most excoriating critiques of industrial capitalism. Dom, who was no socialist, planned to use his conclusion to foreground their radical ideas and more nature-centred cosmologies.

After Raoni, the best known of these voices is Davi Kopenawa Yanomami, whose people dwell in the largest demarcated Indigenous territory in Brazil. He has argued that the term 'climate change' should be redefined as the 'revenge of the Earth'. In the battle for ideas, he has gone on the offensive, with speaking tours of foreign cities, films and books. In his book *The Falling Sky*, he wrote:

> In the past, the white people used to talk about us without our knowledge, and our true words remained hidden in the forest. No one other than us could listen to them. So I started traveling to make the city people hear them. Everywhere I could, I scattered them in their ears, on their paper skins, and in the images of their television. They spread very far from us, and even if we eventually disappear, they will continue to exist far from the forest. No one will be able to erase them. Many white people know them now. Hearing them, they started telling themselves: 'A son of the forest people spoke to us!'

Another of the great Indigenous philosophers and critics of 'white' culture that Dom intended to speak to was Ailton Krenak, whose people were horrendously affected by the collapse of the tailings dam at Mariana. Krenak, who is also a member of the Brazilian Academy of Letters, calls capitalism 'the thing-making machine', which he blames for multiple environmental and spiritual crises. He said Indigenous people are still present in the world because they have escaped this system and learned to live with an almost ever-present apocalypse that dates back to the arrival of the first European colonisers in the Americas. 'The world is slowly, if belatedly, awakening to the fact that Indigenous peoples, who are under threat, have valuable life experiences to share,' he observed in his book *Life is not Useful*. 'Either you hear the voices of all the other beings that inhabit the planet alongside you or you wage war against life on Earth.'

I wonder how Dom would have written about the post-Bolsonaro years. Since he and Bruno died, there have been a few positive signs of change. In 2023, the Workers' Party president Luiz Inácio Lula da Silva promised zero deforestation by the end of the decade, appointed Brazil's first Indigenous minister, Sonia Guajajara, recognised more than half a dozen new Indigenous territories, initiated moves towards a bioeconomy and saw his environment minister, the Amazonian Marina Silva, delay approval for oil exploration near the mouth of the Amazon river and a new licence for the Belo Monte hydroelectric dam. The state once again asserted its presence.

This was progress, though it was uneven and not nearly enough. Indigenous groups in the Javari Valley said they saw no improvement on the ground. Elsewhere, some problems became worse. The agribusiness-dominated Congress moved to limit future land demarcations and tried to push forward new megaprojects, including a major upgrade of the BR-319 highway through one of the last pristine areas of rainforest, a new railway to transport soy from cleared land and new oil exploration close to the mouth of the Amazon. Lula has done little to stand in their way, even actively promoting some of these moves. Like many on the traditional left, his instincts were formed in an age of petrol pumping and concrete pouring. Conservation and climate do not come naturally to him.

It is a reminder, if one were needed, that government-led command-and-control politics are important but limited. Deforestation has been cut by an impressive 50 per cent, but this merely slowed the destruction. The Amazon is still moving ever closer to a tipping point. Global heating has hit record levels. The worst drought in living memory, still ongoing at the time of writing, has pushed river levels to record lows, disrupting supply chains and leading to countless deaths of other species, including more than 200 endangered Amazon river dolphins that perished in the shallow, polluted waters of Lake Tefé in Amazonas state. The tinder-dry conditions created more fuel for forest fires, which flared up with a fury once again in 2024. Indigenous communities, meanwhile, continued to split as traditional culture and values came under assault not just from invasions of miners and loggers, but the spread of homogenising outside ideas through US billionaire Elon Musk's Starlink satellite internet provider, bringing a torrent of fake news to remote villages, as well as positive new possibilities for shared action on dispelling invaders, fighting flames and dealing with health emergencies.

If the forest and its people are to resist this and maintain their diversity, independence and traditional culture, a more profound transformation is needed. There have been encouraging signs in the rise to power of a new generation of Indigenous women politicians, including Sonia Guajajara, Joenia Wapichana, Célia Xakriabá and Juma Xipaya. On the education and policy front, Dom and Bruno's widows have joined the new generation of activists. In the wake of her husband's death, Alé established the Instituto Dom Phillips, which aims to honour his legacy through school exchanges. 'We don't want to be frozen in pain and frustration. We want to forge ahead,' Alé told my *Guardian* colleague Tom Phillips after a trip along the Itaquaí river to visit the monument to her husband and Bruno. 'We must transform this pain into a positive movement – and give new meaning to everything that happened . . . I think that if Dom was here talking to me now, he'd say: "Go Alê: move forwards, learn more, make contacts, help to echo this message about this incredible thing that is the Amazon and all of its beauties."'

Beatriz Matos joined the Brazilian government as head of the Department of Territorial Protection of Isolated and Recently Contacted Indigenous Peoples, within the Indigenous Peoples ministry, where she works with some of the same team as her late husband. The message she is trying to put across is that isolated peoples are important because they have put their relationship with the forest above a relationship with people from the cities: 'Their autonomy is important. It is not that they don't know we exist. They do and they have suffered traumas as a result. Some have almost been wiped out. We must protect their right to that decision (to remain isolated) in and of itself. But we should also be aware that the presence of these people is the reason the Javari Valley Indigenous territory is so enormous . . . There is increasing evidence from archaeology that the Amazon was their garden. Not a garden like those in France, but a garden in the Indigenous style. One that had abundant fruit and areas to hunt. Not based on the domination of one species over another. But a form of living with respect for others. This is not a Disneyland or a fantasy of everyone living in harmony. There are constant struggles. There is death. But it is not a war. It is a form of coexistence with millions of other species . . . It is not just about us. We need to learn how to live with others. To save Amazonia, we need to create conditions for Indigenous people to be themselves. For me, it is essential to work with Indigenous people to do what they have done for tens of thousands of years.'

In this, she is guided by one of the most original thinkers in Brazil, the archaeologist Eduardo Neves, who has upended the accepted wisdom about Indigenous peoples. His ideas revolutionised the thinking of a new generation of archaeologists and gave the world a new perspective on the history of Indigenous civilisations in the Amazon – one that portrays them not just as hunter-gatherers, but as rainforest gardeners and technicians. This is the basis for a completely new way of thinking about the Amazon and its people. Instead of the colonial and military dictatorship-era concept of a sparsely populated wilderness, Neves' reconstruction reveals complex societies that planted natural infrastructure and a diverse forest that has been influenced by humans far more than most people realise. They were highly productive, but not in the destructive, monomaniacal style of modern capitalism. What they produced was more life, more abundance and more diversity.

When the first European colonisers arrived at the start of the sixteenth century, Neves estimates the population of the Amazon was about 10 million and the average life span was higher than that in Europe. There were thousands of villagers and several cities. Trade routes criss-crossed the forest and connected across the Andean mountains with Incas. Over at least 13,000 years inhabiting the forest, these Indigenous groups developed a rich cultural heterogeneity. Every one of the hundreds of different peoples and communities had a different style of pottery and often a distinct language. This continues today, when there are between 180 and 300 languages spoken in the Amazon. Neves believes this was because people valued diversity for its own sake. It was essential to their identity.

He and his team of archaeologists have so far identified 6,000 sites in the Amazon and they continue to find more. It is not easy, nor as obviously glamorous and 'newsworthy' as the classic colonial form of exploration for statues, edifices, ornaments or hoards of jewellery, made famous in films like *Indiana Jones*. The accumulation of wealth and treasure is not a central part of many Amazon belief systems, much to the dismay of the many Western explorers who came looking for hidden cities of gold. In Yanomami culture, for example, when someone dies, all their possessions are burned or discarded. Everything goes back to where it comes from. It is not piled up and passed on to the next generation. So rather than hunting for buried treasure, Amazonian archaeologists must seek signs of human interactions with nature. They trace relationships as much as objects.

The evidence often takes the form of dark soil created by years of cultivation, which can be found in between 2 per cent and 3 per cent of the Amazon. Or it can be inferred by the presence of hyper-dominant plant species, which suggest human intervention in nurturing trees and shrubs that can be used for food, medicine or rituals. Neves estimates that half of all trees in the Amazon bear some connection with Indigenous management practices. 'The very structure of the forest is a consequence of Indigenous intervention over millennia.' This analysis could not be further from the dictatorship-era claims of the Amazon being a 'green inferno' and a 'land without people'. Instead, Neves' theories are a rebuke to the ugly limits of contemporary capitalism. Here in the Amazon was architecture intermeshed with nature; economies based on abundance for all rather than wealth for some; identities centred on diversity and familial relations with an entire ecosystem. These are all radical, highly sophisticated concepts that would have evolved with time and been slightly different from community to community and region to region. They were the opposite of dogmas.

A Life-Changing Relationship: Educate and Rethink

The forest was ever-changing, and so the lives and thinking of its people also had to adapt.

That point often seemed to be overlooked by those who – usually for political or colonial reasons – tended to portray Indigenous culture as backward or locked in time as 'noble savages', and those who romanticised symbols of a pre-industrial, pre-colonial age. In fact, the Amazon's original peoples helped to shape their environments in ways that were more forward-thinking in terms of sustainability than anything that Western economies have managed in the past 200 years. It has only belatedly been acknowledged by science that demarcation of Indigenous territories is the most cost-effective way of storing carbon and protecting biodiversity. That is because forest-dwelling communities have lived as part of nature for thousands of years, and their customs and cosmologies are geared towards nurturing their habitat. Across the world, deforestation and degradation are consistently lower in community-managed forests than unmanaged or unprotected areas.

Which is not to say there is a single Indigenous practice, or that every one of their ideas has to be emulated, or that city people need to return to a hunter-gatherer lifestyle. Nor does it mean Indigenous people cannot embrace new concepts and technologies. The widespread adoption of Starlink satellite internet through the Amazon proves that, as does the proliferation of Indigenous influencers on social media. As well as spreading fake news, social media enables community action. Some environmental activists believe that a combination of Indigenous knowledge and new technology could be the best source of hope for the world's manifold crises. Instead of monetising the forest, they hope modern communications can be used to Amazon-ise the world.

That is still a utopian dream, but it illustrates a broader point about mixing and matching ideas, listening to those who know the forest best and recognising that strength lies in diversity. That very Amazonian message is a healthy and uplifting alternative to the earth-and-soul-destroying trend of monotonising production in the name of economic efficiency. Bruno had come to recognise that. Dom also appeared to be moving in that direction.

There was one last trail to trace. I had followed Dom's precept – 'listen to Indigenous people' – but I still hadn't unearthed what he meant by the notes 'Finishes on the scene in Medicilândia?' and 'Close with the day with the students and professor.' For more than a year, I considered this question without getting anywhere. Then I got a break. The Brazilian journalist Daniel Carmargos told me about a driver who Dom had hired for his epic journey along the Trans-Amazonian highway to visit Anapu, Altamira and far beyond. He gave me his contact details. It turned out the driver and Dom had made a trip together to Medicilândia, in Pará. Antonio Elio Gomes Silva, or Elio as he was widely known, was a schoolteacher who augmented his income by working as a guide and driver for visiting foreign journalists. He remembered Dom with great fondness and spoke animatedly about the trip they had made together in the company of a university professor (who had been taught by Elio as a schoolboy) and three of the professor's students. The solution to the puzzle was now taking shape.

Elio agreed to take me to the places they had visited together. Ahead of the drive, he sent me a link to a video lecture by the professor Anderson Serra and his students Thais Santos Souza, Nayara Souza Dias and Adayciane de Sousa. It was all about their

A Life-Changing Relationship: Educate and Rethink

trip with Dom. They spoke of 'the British journalist Dominic' and the inspiring farmers they had met in Medicilândia, who had developed a sensitive relationship with the forest that offered hope for a more sustainable model of family farming. I started to see why Dom might think this touched on the 'education and new thinking' themes he wanted to address at the end of his book. But what had he made of it?

We set out early on the Trans-Amazonian highway to try to find out. On the way, Elio shared his stories of Dom covering protests by smallholders in Anapu. We discussed religion and the horrifying recent revelation that the killer of environmentalist Chico Mendes had changed his name and reinvented himself as an evangelical pastor in Medicilândia. His true identity had been exposed only when he'd tried to run as a candidate for Jair Bolsonaro's far-right Liberal Party. 'The criminals use religion to launder their reputations,' Elio said. 'And the Evangelicals have become so political. All they care about is access, especially to Indigenous territories.'

Medicilândia announced itself with a giant banner across the road: 'The Capital of *Cacau*.' The city of 27,000 people was named after the most murderous general during the Brazilian dictatorship, Emílio Garrastazu Médici, who was president from 1969 to 1974, when the Amazon was opened up for colonisation. These days, the community prefers to identify more closely with the region's primary produce – cacao, the main ingredient of chocolate and a core element of Mesoamerican culture for thousands of years. As Neves' archaeological finds demonstrated, cacao was domesticated by Indigenous communities in many parts of the Amazon and it has become one of the hyperabundant species that define the forest. Colonising settlers, however, turned cacao into another vehicle of destruction. In the search for higher yields and bigger profits, most of them planted cacao as monocultures that line the roadside in and out of the town with stubby, brightly leaved trees.

Our goal was to meet two farmers who were doing things differently. Darcirio Wronski's plot was located a few minutes after the asphalt of the highway reverted to a dirt track and announced by his wife's brightly coloured sign 'Dona Rosa Chocolate'. After the endless miles of cattle pastures and monocultures, their farm was a refreshingly diverse plot that still resembled a forest. Tall ipê and Brazil nut trees were planted amid the cacao, providing shade and protection against the wind. Darcirio's home snuggled into the vegetation rather than imposing itself on the land. It was a modest building by choice. The family could probably afford a bigger home as the price of cacao had tripled over the previous year and, as one of the first producers in the region to earn organic certification and win international awards for quality, they could charge a hefty premium. But Darcirio, a 75-year-old man with sparkling eyes, said it was never about prestige or money; he had tried to do things differently since he arrived there from the southern state of Paraná in the 1970s.

'As soon as I put my hands in this red earth, I knew this was the place for me. I started with sugar cane but, unlike my neighbours, I didn't clear all the land. I planted trees because I wanted diversity. I learned that from the Kaingang Indigenous people in my hometown in Paraná. I worked at a clinic and spent a lot of time with them. I used to play football with them. I liked what they did. They farmed without fire. They looked after the forest and the water. I told Dom about that when he was here with the professor.'

A Life-Changing Relationship: Educate and Rethink

I could now see a link between 'Listen to Indigenous people' and 'the scene in Medicilândia with the professor and the students.'

Darcirio had also promoted education as the key tool for change. As president of the biggest family farming group, the Association of Casas Familiares Rurais, for the entire northern region of Brazil, encompassing several Amazon states, he created a network of training institutes that sent young people to learn at farms that innovated and experimented. It had nurtured and inspired many to try biodiverse agroforestry.

It had only partially caught on. The majority of settlers still preferred monocultures or cattle ranches, both of which depended on forest clearance by fire. Darcirio lamented the consequences. He said had could feel the change in the climate. 'We used to have silent rain all through the year, now we have long dry periods and when the rain comes, it is noisy. It comes with wind and thunder,' he said ruefully. 'I don't understand the minds of some people. The destruction of the Amazon is one of the greatest crimes in the history of the world. It is better to have a healthy forest and a good climate. That makes you feel happier and purer. Planting is life.'

His closest ally in the region was our final port of call. José Osmar Couto, or 'Zé Gaúcho', as he was best known, was another pioneer of agroforestry who Dom described in his notes as 'the guy who went his own way'. He too had travelled north to the Amazon on a government promise of free land, driving for 30 days from Porto Allegre in 1971 with two friends in a truck that now sits rusting on his plot of 276 hectares, 45 per cent of which was forest while the rest was cacao mixed with taller trees. He claimed to have been the first in the region to halt the use of pesticides and he resisted a government suggestion to mix the cacao with a non-native species, gmelina. Instead, he opted for mahogany, even though he knew it was protected and could never be harvested.

At 81 years old, he took me through his land. By not using chemicals, by protecting his water sources and introducing a diverse agroforest technique, Zé Gaucho said he had shown it was possible to thrive as a farmer and do the right thing. The professor Anderson Serra lauded this example to his students. And Dom had clearly been impressed. In Zé Gaucho's visitor book was my old friend's familiar scrawl: '28/8/2021 Dom Phillips, *Journalista Ingles. Linda floresta que Sr. cresceu.*' (Beautiful forest that the *senhor* has grown).

I was moved to find Dom's handwriting in that book. Reading it, I was reminded of Dom's '*Amazônia, sua linda*' post before his death. I could imagine how encouraged he might have felt to see such a positive example – not just for himself, but through the eyes of the young students who were there with the professor. Even amid the vast destruction of Medicilândia and the entire Trans-Amazonian highway region, here was proof that it was possible to do things differently if people opened their minds, if they listened to Indigenous people. Then, maybe, instead of a war with nature, there could be a relationship.

That seemed to be what Dom wanted this book to achieve. As he wrote in his proposal: 'As the reader gets to know the people who are living with the immediate consequences of local and international policy in the Amazon, they will see how those people's lives interlink with ours, how the fires that blaze and the trees that fall in the rainforest affect the whole planet, and how decisions we make at local and national level in our own countries can, genuinely, help to save the Amazon.'

A Life-Changing Relationship: Educate and Rethink

The core Amazon battle was for hearts and minds. Sure, defending territory on the ground was a crucial first step. Securing government support could then slow destruction. Bringing transparency to beef and soy supply chains would help too. As would a rethink about destructive infrastructure projects. Valuing forests more alive than dead would be a game changer. Securing international finance should accelerate the transition to a sustainable future. Ecotourism and carbon taxes might have a role to play. Compelling global pharmaceutical companies to share the benefits of biodiversity would incentivise conservation and support livelihoods. But for all of these ideas to work, what mattered most was a healthier way of thinking about the forest. It was about listening, about building a new relationship with nature. Or, better still, rediscovering the virtues of an old one. That did not need to be complicated. It could be instinctive. It could be the feeling of delight in seeing the world as it should be. It might even start with a simple expression of joy: *Amazônia, sua linda!*

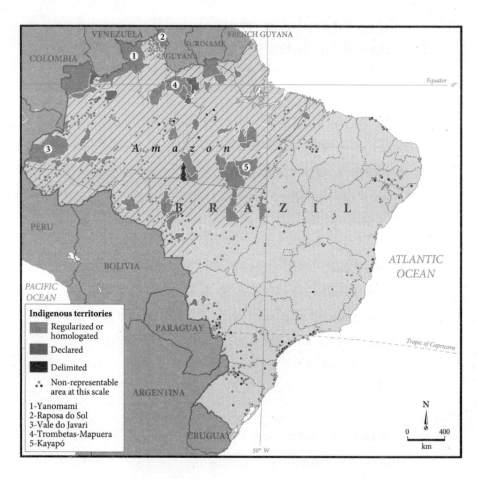

Recognised Indigenous territories

AFTERWORD

Listen to the Forest: Indigenous inspiration

Beto Marubo with Helena Palmquist
(Translated by Julia Sanches and Diane Whitty)

Beto Marubo was born in the Javari Valley. Since early 2018, he has represented the Indigenous peoples of the Javari in their dealings with state institutions. For more than two decades, he has also worked as a coordinator of the ethno-environmental protection programmes designed to protect peoples who live in voluntary isolation.

Helena Palmquist is an Amazonian writer and activist, born in Roraima and raised in Belém, who has worked on Indigenous and environmental issues throughout her career as a journalist and anthropologist, both as a public servant in the judicial branch and as an advocate for the rights of isolated Indigenous peoples.

'It is not external experts who have the answer but the people who have nurtured and protected these forests for millennia.'

'That's presumptuous.'

These are the words I said to Dom Phillips when he first told me he was planning to write a book on how to save the Amazon.

As an Indigenous man raised in the rainforest, I was sceptical. 'But you're a foreigner,' I told him.

The reason for my cynicism is that the forest has had enough of white 'saviours'. So many city academics and experts come here only to go back home and write books for their people. Those books do us no good. In fact, sometimes they make things worse. Scholars and experts need to change how they talk to people. They need to help us convey our own centuries-old knowledge of the forest.

As I got to know Dom, I realised he was different. Humble, honest. He said he wanted to listen to those who have been protecting, managing and planting the forest for millennia: Indigenous peoples, *quilombolas*, *ribeirinhos*, members of social and environmental movements, native Amazonians.

Dom had already met my friend Bruno Pereira. 'He looks after Indigenous people there like they're his brothers,' Dom said.

I considered Bruno Pereira to be my 'white brother'. He was a methodical, careful scholar of the Indigenous world, but most of all he loved being in the forest and living alongside its people. This made him a person of rare sensibility in Brazil because we still haven't managed to raise environmental awareness in our own society, and not even all Amazonians are convinced of the vital role played by forests. It seems a bit twisted to me that we need people like Bruno and Dom to persuade Brazilians that our future depends on the Amazon.

My people are the Marubo, one of seven ethnic groups who inhabit the Javari Valley. In my community, they call me Wino Këayshëni. I come from an area near the headwaters of the Curuçá,

Listen to the Forest: Indigenous Inspiration

a major river running through the Javari. My people have lived in this region for centuries, as our history tells us. My village, Maronal, is one of 59 Indigenous communities in the Javari Valley Indigenous Territory, which is also home to the Marubo, Mayoruna, Matis, Kanamary, and Kulina (Pano) peoples, as well as to the Korubo and Tsohón-Djapá, two recently contacted ethnic groups.

Along with these seven peoples who engage with the surrounding society, there are 16 groups who have no ongoing interactions with the non-Indigenous world. Known as isolated peoples, they live according to their ancestral customs and thus rely on forest ecosystems. They source all of their food, water, shelter, tools, adornments and medicine directly from the rainforest. They are amazingly skilled at orienting themselves in the woods and can travel long distances by following an internal GPS many of us have lost. All Indigenous peoples in the Amazon once lived like this. My land is unique and crucial to the survival of these isolated peoples because, according to official data, it harbours the planet's largest concentration of groups living in voluntary isolation.

I eventually became a spokesperson for the rights of my parentes (as Indigenous people refer to each other in Brazil) and the isolated peoples living in the Javari Valley. Since early 2018, based on a decision by peoples in the area, I have represented the Indigenous movement outside our territory, conveying the interests and demands of these peoples to the state institutions that influence daily life in our communities. Earlier, I worked as a coordinator of two ethno-environmental protection programmes, which are units of Brazil's Indigenous affairs agency, FUNAI, designed to protect isolated peoples. I have been involved in this cause for more than 20 years. Over the centuries, these peoples have opted to preserve their autonomy, a decision that leaves them extremely vulnerable. Oftentimes, the choice to continue living in isolation, in remote, hard-to-reach regions, has been reinforced by some traumatic contact in the past.

It is thanks to them that we Indigenous peoples of the Javari inhabit Brazil's second-largest Indigenous territory, which figures among the largest intact tropical rainforest reserves on Earth. In our territory we find refuge from disease, poverty and racism, which have characterised Indigenous peoples' experiences with Brazilian society since contact. This is why I believe the rights of these isolated groups and the protection of all Indigenous territories in Brazil are essential to the survival of our forests and, ultimately, of the planet.

I've seen an isolated group's first contact with the outside world. Now and then, they come to talk to me. I remember an elderly man from the isolated Korubos; after decades without any contact, he and his family approached us. The old man, Pëshkén, came up to me and asked what those things were that made noise in the sky. He said they used to hear this racket above their village when he was younger, but now his people were hearing it more and more often: a 'powerful whirlpool' overhead. I tried to explain it was planes flying over the valley, but he couldn't process the information. So I came up with a comparison more relatable to his everyday life, and said they were 'flying canoes'. He looked at me incredulously when I explained that the canoes were taking white people to distant lands and that many of the craft were carrying more people than all the inhabitants of his village. Pëshkén told me he was worried because white men kept getting closer to them. He said they regularly found traces of hunters and fishers. The elderly man seemed confused, unsure about how to react. He asked me, 'Who are these white men? Are they good men?' And I had to tell him, 'No, they're not. They're bringing the end of the world.'

For almost 500 years, Brazil's policy toward Indigenous people was one of forced contact. Brazilian society saw entire peoples vanish until, one generation ago, we decided we'd seen enough Indigenous death. So we began respecting isolated people's right to self-determination, their right to avoid contact with others, and we started working for territorial protection on their behalf. When Brazil enacted its Constitution of 1988, following the demise of the military dictatorship, this right became state policy.

This new approach to Indigenous rights, adopted by Brazil in the 1980s, has been embraced by several other Amazonian countries as well as the international community, which has also come to acknowledge the role of our people in conserving ecosystems. Today, Indigenous territories are recognised as the finest examples of forest protection, pivotal to maintaining climate balance in Brazil and on the planet and to preventing the biodiversity loss that also threatens us. We, Indigenous peoples, defend these forests with our very lives because, for us, they are life.

I learned this first-hand as a child, when I lived with my elders in the Javari. They were always expressing concern about our mother, talking about how important she was, Earth and every being who lives on her – what outsiders call biodiversity. Our daily lives in the village have always reflected this. Older people would tell the younger ones: when you grow up, think about your grandchildren, who will need there to be peccaries (*Tayassu pecari*, an

Amazonian pig) in the forest for them to hunt. We don't have grocery stores like the *nawa* (as non-Indigenous people are called in the Pano languages of the Javari Valley).

These were words of caution to us, enthusiastic young people who were learning to hunt at a time when guns were replacing arrows. Since ammunition was expensive, the younger ones among us were given arrows to practise with, making the kill much harder. Only three out of every ten arrows would hit the target, especially when we were hunting monkeys. It was a time when the valley had a lot of birds like jacus, tinamous, curassows (prized for roasting). With a gun, one shot was enough. Then, around the 1990s, our elders got worried and started telling us it was important to hunt only what we needed in order to protect the land, waters, rocks, forest and animals – what the scientists on the outside call ecosystems.

Our elders always cautioned us not to kill mothers and babies when hunting peccaries. Only the ones that were older and bigger were to be slaughtered. Little ones, mothers – never. The same with tapirs. The elders took this very seriously and taught it to younger generations. Women followed the same logic when tending plants. Our aunts would harvest food, and we youngsters would carry bunches of bananas and cassava. And what our elders said about animals, the women said about harvesting. They told us not to cut buriti palms because we needed them to extract a fermented drink. If we cut them, then they would die. The fruit should be harvested, leaving the trees intact. Just as we did with açaí, so that there would always be more. Tucum palm trees, which are used to make hammocks, are harder to deal with. You have to climb two trees at once to get to the top and remove the young fibres, which are better for weaving. It would be much easier to cut down the trees but then the tucum would be gone.

We Indigenous people don't see land, water, animals and plants as resources. Land and water are mother and father to the Indigenous peoples of the Amazon, the cradle that births and sustains life. That land could be owned by a single person is unthinkable to us because everyone's survival depends on it. We can't imagine how anyone could pollute, divert or dam water for their own profit. We hunt and eat animals but don't consider them inferior or less intelligent. We believe plants are wise beings who provide us with shelter, food, medicine and joy, and often bear witness to our passage through life, like a samaúma tree that watches several generations of Indigenous children play among its majestic roots over the course of its 100- to 200-year life.

This isn't how the *nawa* think. They look at a tree and ask themselves how many boards they can tear from it. They know nothing about planting gardens where different plants help each other grow, which also makes us healthier. They want to plant a single crop on farms, killing the soil to reap money instead of food. When they hunt, they pile up dozens of animal bodies of the same species to sell and make more profit. They forge documents declaring themselves the owners of land they've never set foot on. They look at the Amazon's mighty rivers and think about all the money they could make by diverting them from their courses.

This is why the dilemmas we face in the Amazon and the Javari Valley are more serious than ever. Organised crime is encroaching on our territory, often invading it to steal timber and poach animals, set up destructive gold mines, fish illegally. This is why Bruno and Dom were brutally murdered. This tragedy, where we lost two people so important to our country and the Amazon, bears the fingerprints of organised crime, whose ranks include politicians and security forces.

It's an issue of time and scale. The *nawa* will always want scale, a larger volume of resources so they can make a profit. This is why Bruno fought illegal fishing. He saw boats leaving our territory carrying tons of *pirarucu* (an Amazonian fish) and felt the need to contain the destruction. Time is also important because what a village consumes in one year is not the same as what a city consumes in the same time span.

On the issue of the Amazon, we need to be very careful to avoid the rhetorical trap that claims Indigenous people need money and need to be productive. This is the discourse of the far right, of ruralists and large landowners, discourse that is used to sell Indigenous land, or what they call 'leasing' these days. It means justifying the destruction of land by claiming Indigenous people need money. We mustn't fall for it. What we need is good public policy. The authorities need to listen to the few organisations that really know the Amazon. Contrary to what was said under Bolsonaro's government, NGOs know the region very well. They're in the forest, with the forest dwellers.

FUNAI used to act as this intermediary. But after the agency's restructuring in 2009, when they abolished job posts on Indigenous land, nearly every employee was removed from the field and transferred to the city. The only ones left in the forest were the NGOs and us, Indigenous people and *ribeirinhos*. We see solutions being proposed but what we don't see are ideas coming from people who live in the Amazon. Solutions are imposed from the outside. This was what Dom Phillips found concerning. As do I.

Listen to the Forest: Indigenous Inspiration

You arrive at a very isolated community and find someone from an NGO or social movement drinking *caxiri* or taking *rapé* with Indigenous people. What you don't find is anyone from the government in charge of planning, development or the economy. You spend ten days in a canoe and meet people from the village's civil society, but never anyone from the government. And the government comes up with some of the worst, craziest ideas – because there's a new trend, because some European economist or someone from the market thinks it's interesting. Even carbon credits, which are often questioned, could be attractive if they involved local consultation and participatory control. But this is not what we've seen happening.

The carbon cowboys arrive in helicopters offering surreal amounts of money and asking Indigenous leaders to sign on the dotted line. We're talking millions of dollars. And our struggling kin are practically coerced to sign. European companies are already in Javari Valley, where they are jeopardising the autonomy of Indigenous peoples. How can we ask isolated communities how they feel about participating in the carbon market? If we don't thoroughly rethink this proposal, it will become yet another issue that sparks conflict in Indigenous villages and lands – just like money from illegal gold mining and fishing or construction of hydropower plants like Belo Monte.

We need representation. But politicians are under the thumb of powerful agricultural and mining interests. Brazil is ruled by congresspeople who live in a bubble, somewhere that exists only in their heads. They want to raise more cattle, plant more soybeans and destroy the largest forest on the planet just so they can make more money. To do this, they are trying to kill us Indigenous people, to push us off our land.

Right now, we see this happening again. The biggest political challenge of my generation of Indigenous leaders in Brazil is the so-called Marco Temporal, or cut-off point, a legal artifice devised by large landowners and Congress that would drastically curb our right to territory. According to this argument, the only Indigenous people who have a right to their land are the ones who were living in their territory at the time the Constitution was enacted in October 1988. This is historically unjust and ethically wrong because many Indigenous people were violently expelled from their land before that moment.

Lula's government, which we trust and support, has not shown the courage required to combat this offensive.

We see this same lack of courage in international climate negotiations. The Brazilian government talks about clean energy abroad but back in Brazil, the state-owned Petrobras announced it will begin exploring for oil near the mouth of the Amazon.

Many people don't know this, but during the dictatorship, Petrobras perpetrated crimes against isolated peoples in the Javari Valley. Petrobras's actions in the region in the 1970s and 80s were an environmental and human disaster. There were deaths. Petrobras used FUNAI to disperse then-isolated groups, attacking and killing them. East of our territory, there are still oil wells that were drilled during the military regime. To this day, some of the mud in the area is green from mining chemicals. Fifty years later!

The people in charge of my country insist on recommending projects we know will be new vectors of devastation, like the paving of BR-319, a highway built during the dictatorship that penetrated the most well-preserved parts of Amazonas state, where my territory is located. Another project that Indigenous people are fighting is Ferrogrão, a railway that would slice through the forest in Mato Grosso and Pará, regions home to Indigenous people we've never even met. We know, based on past colonialist projects of a similar nature, that these undertakings will lead to an uncontrollable spike in deforestation.

Nawa scientists have finally realised that these actions are pushing the Amazon to a point of no return, one where the forest can no longer be saved because the soil will not have the strength to regenerate trees and everything will become dry. Shamans and Indigenous leaders have been warning people about this for a long time. And they know that what is bad for the Amazon is bad for the world. Their climate fears are fast becoming our reality.

It seems my country is incapable of seeking new solutions. Instead, it insists on pursuing an idea of development in the Amazon that not only never worked but was responsible for the genocide of various Indigenous peoples. I get the impression that there is a drawer in Brasília filled with a backlog of projects from the dictatorship and that each democratic government opens it whenever they want to draw up plans for the region, whether it's the paving of BR-319, another hydropower plant, oil exploration or railways. It's all old news that keeps cropping up again and again.

On a more positive note, I've been thinking about the creation of the Ministry of Indigenous People during Lula's third term. It's an opportunity for Indigenous communities to finally advance policy, something no government has ever given us before. But, in a way, the government is also co-opting the Indigenous movement. The presence of Indigenous people is useful to Brazil's foreign policy, to the trade agreements it reaches in Mercosur or with the

European Union, agreements that are often not in the best interest of the Amazon or its people.

I'm still trying to answer Dom's question: how to save the Amazon? For me, we have to start by recognising a fundamental truth: there isn't just one Amazon but many. We must internalise this truth and think about the specificities of each place. The public policies needed in my village, Maronal, which lies ten days away from Atalaia do Norte by boat, are not the same ones needed in villages much closer to the city.

No Indigenous person, regardless of what forest they inhabit, would ever consider destroying their ancestors' spaces, the places where their ancestors died and were buried. I am Marubo but live in Brasília, and I always miss the river where I was born. All Indigenous people who, like me, live far from their territory, still make plans to be buried on their land. We need to return to the earth, to community. Which is why it is unimaginable that we would ever destroy that space. A territory is not like earth. It is more like a paradise you return to for good.

It should be people of the Amazon with this kind of vision, people living in the forest itself, who make public policy for the Amazon. Now and then I attend UN meetings in Geneva or New York City as a representative of Brazilian Indigenous peoples. I've seen a lot more external experts there than residents of the Amazon.

All the signs, the data, show us that the parts of the Amazon where the forest is truly protected lie on Indigenous land. This is why we must demarcate as much Indigenous land as possible, as quickly as possible. But what's happening is the opposite. Despite the promises made by the Brazilian Constitution, the government has been delaying demarcations and setting malicious timeframes.

We're well acquainted with the logic of the *nawa* market, which claims that agribusiness and mining are important for the country's GDP. But how much does it cost the GDP to cause environmental disasters and jeopardise our future? The numbers don't add up.

As much as Brazilian society watches increasingly serious and deadly climate catastrophes on TV, they still don't seem to understand that these disasters are being provoked by the greed of agribusiness, mining companies and the congresspeople who represent their interests. How can we get society to understand that rivers are being poisoned, forests devastated and the climate turned upside down because they are choosing agribusiness and mining?

Demarcating territories – Indigenous but also *quilombola* and *ribeirinho* – is a matter of protecting the future for all and also of protecting human rights. It is a matter of Brazil having the integrity to protect the territories of people who have chosen a way of life that is essential to the country's future. To protect isolated peoples who live this way because they've long survived massacres and environmental devastation. We must protect them. It's the bare minimum.

I'd like to see a sincere effort made in Brazil, because without the Amazon and its people, climate catastrophes will become more frequent, not only in our country but also abroad. I'd like to see the rest of the world, which claims to be concerned with the longevity of our forest, push for stronger Indigenous and environmental agencies. Wealthy countries should finance these efforts and also help us prevent the Brazilian government from adopting ruinous and outdated economic solutions that often benefit only European and US companies.

I would like to see government authorities make a real effort to understand Indigenous people and actually be in the forest with us. It doesn't make sense to only send military or police intelligence into the forest for special operations. Those who formulate policies are the ones who need to spend time with us there. They must shake off their non-Indigenous arrogance and listen to us, eat wild game and go fishing with us, and sing our songs with us in the forest.

This is the lesson Bruno Pereira and Dom Phillips have left Brazil and the world. They were in the forest with Indigenous people, walking, eating and singing alongside us. Bruno, with his unforgettable laughter and impatient courage, had enormous admiration for the Indigenous leaders he worked with and made a point of learning from them every step of the way. Dom, sweet and attentive, always heeded what they told him and showed him in the forest. Simply remembering them makes me believe the Amazon can still be saved.

Dom wanted to do more than just write a story. He wanted to help. To truly help. We need more Doms, more Brunos. Both came to us so they could see our reality and listen to what we said. And both did what they could, in their own way, to respond. They were brave and they acted. If everyone did the same, we might begin seeing change.

A final point: it isn't enough to Indigenise politics. We also need ideas, knowledge and wisdom to come from the villages, *quilombos* and riverbanks. It is not external experts who have the answer but the people who have sown and protected these forests for millennia. The Amazon mustn't be viewed as a problem; it is the solution.

ACKNOWLEDGEMENTS

Dom Phillips and Tom Hennigan

Among the papers for this book that Dom had accumulated by the time of his death was a short list of names under the title 'Acknowledgements'. Like the rest of the book, we can be sure this list was a work in progress, an aide-mémoire to be rounded out and written up into something similar to the generous acknowledgements that the reader encounters in his first book, *Superstar DJs Here We Go*. How he would have done so we cannot know and so the main objective here is to thank the many people who were determined that Dom's murder would not see him silenced and worked to ensure that *How to Save the Amazon* was completed and brought to publication.

Following the events of June 2022, as conversations turned to the fate of his book, Dom's widow Alessandra asked a group of his journalist friends to explore the possibility of completing it. An editorial steering group was put in place headed up by Dom's old friend from their Rio days together, *The Guardian*'s global environment editor Jonathan Watts, who has since taken up residence in the Amazon rainforest municipality of Altamira. Another friend was Andrew Fishman, the Rio-based president of the Intercept Brasil, who had engaged in deep debate with Dom about his original idea. From London, they were joined by his agent Rebecca Carter, a key supporter whose early belief in the book had secured a publishing contract and the all-important advance that allowed him to proceed. Also participating from London was David Davies, Dom's friend, colleague and confidante since their days working together at *Mixmag* back in the 1990s, and Tom Hennigan, his friend from São Paulo days and Latin America correspondent for the *Irish Times*.

In the words of Dom's sister Sian, the task of completing *How to Save the Amazon* became 'not only a way to create a lasting tribute, and give meaning to his death, but also a contribution to the urgent efforts to find solutions to the crisis in the Amazon'. For the editorial group, the challenge of finishing it created a means of channelling the grief and anger so many of Dom's journalist friends felt at his murder into a unique act of solidarity with a much-admired colleague. This book is Dom's but many others had a part in making sure that, though incomplete at the time of his death, it did not die with him.

Alessandra and the editorial team would particularly like to thank those colleagues who, alongside Jonathan and Andrew, volunteered to take on the chapters Dom had yet to write, often retracing his steps as they searched for answers to the question he posed in his upbeat original title: *How to Save the Amazon: Ask the People Who Know*. In Altamira, Brazilian journalist Eliane Brum, founder of the environmental website Sumaúma. In Rio, British journalist Tom Phillips, *The Guardian*'s Latin America correspondent. In Costa Rica, US writer Stuart Grudgings, another of Dom's former colleagues from their time in Rio de Janeiro. Also in New York, US journalist Jon Lee Anderson of the *New Yorker*. Thanks also to Beto Marubo and Helena Palmquist, respectively coordinator with the Union of Indigenous Peoples of the Javari Valley (Univaja) and adviser to the Observatory of Human Rights of Uncontacted and Recently Contacted Indigenous Peoples (Opi), for the afterword.

Acknowledgements

Thank you also to photographers Gary Calton, João Laet, Nicoló Lanfranchi, Lianne Milton and Fábio Erdos, friends and colleagues of Dom who joined him on reporting trips in the Amazon and whose work appears in these pages. Also to Pedro Biava and John Mitchell, Dom's old friend and colleague from their Bristol days together, for sharing their images. Finally, grateful acknowledgement to Agence France-Presse and the audiovisual team at the Palácio do Planalto in Brasília for permission to reproduce their photographs.

Many of Dom's colleagues and friends from Rio and São Paulo volunteered to help with this project. Our gratitude to them all, especially those who read and helped to edit chapters in manuscript: Claudio Angelo, Brazilian journalist and author (with Tasso Azevedo) of O *Silêncio da Motosserra* (*The Silence of the Chainsaw*). Ana Aranha, Brazilian journalist and documentary filmmaker with *Repórter Brasil*. Paulo Barreto, founder of the Brazilian research institute Imazon. Vincent Bevins, US author and former colleague of Dom's in São Paulo. David Biller, US journalist and Brazil news director for Associated Press in Rio de Janeiro. Kátia Brasil, Brazilian journalist and a founder of the Amazônia Real news agency. Sonia Bridi, Brazilian journalist with the Globo network. Gareth Chetwynd, British journalist and old friend of Dom's from their Rio days. Daniel Camargos, Brazilian journalist with Repórter Brasil and another colleague of Dom's on Amazon reporting trips. Sylvia Colombo, Brazilian journalist and one of Dom's first local colleagues after his arrival in São Paulo. Sam Cowie, a British journalist based in São Paulo, where he first met Dom. Otavio Cury, Brazilian filmmaker and one of Dom's first friends in Brazil. Wyre Davies, Welsh journalist and former South America correspondent for the BBC.

Andrew Downie, Scottish author of *Doctor Socrates* and Dom's friend since their São Paulo days. Thomas Fischermann, German journalist and former Brazil correspondent for Hamburg newspaper *Die Zeit*. Fabiano Maisonnave, Brazilian journalist and Amazon correspondent for Associated Press. Canadian journalist Stephanie Nolen, now the global health reporter for *The New York Times*, Rubens Valente, Brazilian journalist with Agência Pública and author of *Os Fuzis e as Flechas* (*The Rifles and the Arrows*). Important contributions were made by Andrew Wasley, a journalist specialising in environment issues at the Bureau of Investigative Journalism. Bibi van der Zee, editor on *The Guardian*'s environment desk.

Social media campaigning, fundraising and moral support also came from Ali Rocha, freelance journalist; Dom's old London friend Clare Handford, TV producer; Tariq Panja of *The New York Times*; Jan Rocha, author of multiple books on Brazil; Vinod Sreeharsha of the *Miami Herald*; Bruce Douglas, former Rio-based freelancer; Lucy Jordan of Unearthed; Julia Dias Carneiro of BBC Brazil; Katy Watson, former BBC South America correspondent; Scott Wallace, author and University of Connecticut assistant professor; Daniela Chiaretti, Brazilian journalist and environmental correspondent with São Paulo newspaper *Valor Econômico*; Simon Romero, US journalist and *The New York Times* correspondent in Mexico City; Douglas Engle, Rio-based freelance photographer; Philip Reeves, international correspondent on National Public Radio 2004–24; Adele Smith and Sebastian Smith of AFP; Maximo Anderson, researcher; Taylor Barnes, a field reporter at Inkstick Media; Anna J Kaiser of Bloomberg News; Catherine Osborn, journalist and foreign policy columnist; Kate Steiker-Ginzberg, attorney currently with the ACLU-Pennsylvania; Nadia Sussman, video journalist, ProPublica.

Acknowledgements

Thank you to *The Guardian*, especially Kath Viner and Natalie Hanman, for keeping this story in the news.

The editorial group would also like to thank Stella Giatrakou Sarah Braybrooke, Liz Marvin, Leonie Lock, Justine Taylor, Jane Rogers and everyone at Bonnier for their commitment to seeing the book published under such difficult circumstances; also Margaret Stead and Ruth Logan, who were responsible for acquiring Dom's proposal for publication by Bonnier. Thanks also to Sian Phillips and John Mitchell, who waded through Dom's notebooks transcribing his scrawl to help out chapter writers, and to Roberta Mello for digitising every single page. Also to Julia Sanches and Diane Whitty for translating the chapter 'A Cemetery of Trees' and the afterword from Portuguese into English. And to assiduous fact checkers Plinio Pereira Lopes, Gustavo Queiroz and Douglas Maia.

Dom's initial project was made possible by his receipt of an Alicia Patterson Foundation Fellowship in 2021. The effort to complete it after his murder was then helped by a Whiting creative non-fiction grant in 2023, the first time the award has been given to a collaborative project. Other crucial support came in the form of a generous grant from the Fund for Investigative Journalism. Dom's family are also deeply grateful to Teresa Bracher, another early backer of the attempt to see this book finished. Alessandra and the editors are also hugely grateful for the fundraising efforts that took place after the murders of Dom and Bruno and which played a decisive role in seeing the book finished, particularly the efforts of Dom's niece, Domonique Davies, her mother Helen and Dom's nephew Caden Phillips. Thanks too to those who organised fund-raising events, including all at the El Sueno Existe Festival, in Machynlleth, Wales. Hundreds of the donors are listed below, but many more provided financial support anonymously.

Thanks to Fiona Frank and Alison Cahn as well as Dom's brother-in-law Paul Sherwood in Lancaster. Also to Dom's former friends from the electronic music scene. Following his murder, Frank Broughton, David Davies, Frank Tope, Bill Brewster, Andy Pemberton, Jerry Perkins, Dan Prince and Dom's many *Mixmag* friends were quick to organise a fundraising event to finance this book, featuring sets from notable DJs Darren Emerson, Jon Carter, Bill Brewster, Dave Seaman, James Lavelle, Lottie, Chad Jackson and DJ Gruff. Likewise, Ben Turner and the International Music Summit also played a big part in making this book possible with their charity auction. Also to Julia Carneiro, Andrew Fishman, Kate Steiker-Ginzberg and Jonathan Watts for organising an online crowdfunding effort. Most of all, thank you to everyone who collaborated in and contributed to these grassroots efforts. Your solidarity with Dom sustained the effort to see this book through to publication.

For help in the search for Dom and Bruno a special thank you to the members of Univaja, Orlando Possuelo and others in Atalaia do Norte. Also to Henrique Cury, Dom's friend who first invited him to Brazil. The editorial group sincerely thanks the families of Bruno and Dom for their support and trust, especially Alessandra, Dom's sister and brother, Sian and Gareth, and brother-in-law Paul.

As found in Dom's papers after his murder:
ACKNOWLEDGMENTS
Alê
Rebecca Carter
David Davies and Martina Klett-Davies
Andrew Fishman and Cecília Olliveira
Philip Reeves
Richard Lapper

Acknowledgements

Margaret Engel
Margaret Stead
Marcos Wesley
João Apiwtxa
Sian, Paul, Gareth
Dean Belcher
João Laet
Daniel Camargos
Tom Phillips, Bibi van der Zee, Martin Hodgson,
    Alan Evans, Tracy McVeigh

The donors to the crowdfunding campaign that helped make the book possible:

Adam Bennett
Adam Mekrut
Adrian G Allan
Adriele Marchesini
Aine Shannon
Ali Hill
Alice Goodman
Alice Mayer
Alison Cox
Alison Tyrell
Alyce Dodge
Ana Ionova
Andie Nesbitt
Andrea Troncoso
Andrew Cowie
Andrew Fletcher
Andrew Naylor
Andrew Revkin
Andrew Richards
Andrew Thomas
Andy Hornby
Angel Figueroa
Angela Frewin
Angela Shaw
Ann Hitchens
Anna Lenk
Anna Pack
Anna Robertson
Anne Louisa Casement
Anne Wooding
Anne Wyatt
Annie Lawler

Antoine Robin Ltd
Antonia Bovis
April Knowles
Arlene Washburn
Arturo De Frias
Aurea Garibaldi
Ayla Bedri
Barbara Covey
Ben McCabe
Ben Nohr
Ben Pearcy
Ben Sadek
Bernardo Maranhao
Blanche Rowen
Bob Frith
Bob Jamieson
Brazil Matters
Brenda Donovan
Brian Edwards
Brian O'Connor
Brian Webster
Bruce Douglas
Bruno Araujo
Bruno D'Acri Soares
Bruno Travers
Carol Arnold
Carol Turner
Carole Bishop
Caroline Wood
Caroline Yapp
Carrie Sandahl
Catherine Luse
Catherine Schwartzstein
Caz Royds
Ceri Mumford
Chantal Adele Smith
Christian Daniel
Christine Ramp-Wolf
Christopher P S Klinger
Ciara Gray-Shannon
Clare Birks
Clare Downs
Clare Handford
Clare Rose
Claudeline Louis
Claudia Miranda Rodrigues
Claudia Quinonez
Clelio Rocha
Constance Malleret
Cora Tudor
Courtney A Crumpler
D A Mendelsohn
D T A Mitchell
Damian Mould
Damian Platt
Daniel Collyns
Daniel Ribble
Daniel Swanson

Acknowledgements

Darryl Fong
David Biller
David Brock
David Davies
David Garratty
Debora Gouvei
Deborah Hofman
Derek Price
Diane Bell
Dr. E Heath
E Hill
Edward Davey
Eileen Freeman
Elaine Lee
Elizabeth Bell
Elizabeth Heaphy
Elizabeth Milton
Elizabeth Slocum
Ella Sprung
Ellen Punyon
Emma Watts
Erika Berenguer
Evelyn Escatiola
Filomena Di Stasio
Fiona Haslam
Frances O'Rourke
Frances Watt
Francis Mcdonagh
Fred L Edwards

Gabriel Funari
Gareth Chetwynd
Gareth Morgan
Gareth Phillips
Gary Calton
Gavin Marwick
Gavin Smith
Georg Schäff
George V
Giles Hayward
Gill Jennison
Gillian Wallington
Giselle Letchworth
Glyn Phillips
Graham Luke
Gregory Gludt
Guy Edwards
Guy Shrubsole
Gwendolyn Knox
Hannah Mullaney
Hannu Toropainen
Hans Kainz
Heidi and Mike Gibbs
Helen Armstrong
Helen Beare
Helen Davies
Helen Fry
Helen Lord
Helen Stevens

Henrik Jonsson
Henrique Terra Lima
Hilary Furlong
Hilary Tyrrell
Ian Buckley
Ian Carney
Ian Vincent Waldron
Ismene Brown
Ivan Nunes
J Da Rocha
J Harrison
J Hudson
J Williamson
Jack Nicas
Jacqueline Power
Jake Wallington
James Andrew Shelton
James Durham
James Haigh
James Milligan
James Savage
James Schumacher
Jan Royle
Jane Macdonald
Jane Thorne
Janet Davies
Janet Swan
Janice Stott
Janis Kershaw
Jathan Rayner
Jean Mills
Jeanette Sharp
Jenny Hoy
Jeremy Kynaston
Jessica Smeall
Jill Gregory
Jo Caryl
Jo Jenkins
Jo May
Joanna Powell
Joanna Service
Joanne Rippin
João Telésforo Medeiros Filho
John Bacon
John Mcclean
John Mitchell
John Weyman
Jon and Anne King
Jose Pedro De Oliveira
Joseph Murphy
Joseph Patel
Joshua Berger
JP Connolly
Judith Kaluzny
Judith Wildman
Julia Blunck
Julia Hall
Julian Caldecott

Acknowledgements

Julie Mccann
June Arthur
K Walsh
Karen Bell
Karen Rawlinson
Karen Yarnell
Katherine Mcnulty
Kathleen Martin
Kelly Caldwell
Kristy Poulton
Lauran Emerson
Laurel Swift
Laurie Blair
Lawrence Jones
Lee Willocks
Leonor Grave
Liam McAllan
Linda Hodgins
Linda Wilhelm
Livia Serpa
Liz Baker
Lorraine Wulfe
Lotte Kehlet
Louise Benson
Louise Bonney
Lucy Jordan
Luisa Piette
Luke Davis
Lydia Duddington
Lynn Nikkanen
M Fatima Carvalho
M P Feehily
Mandy Greenwood
Mandy Knott
Marcela Olavo Leite
Marcia Reverdosa
Marcus Wright
Margaret Hall
Margaret M Iggulden
Maria De Fátima Costa
Maria Luiza Nery
Marjon Van Royen
Mark De Rond
Mark Harris
Mark Leonard
Mark Rennie
Mark Williams
Martin Gugg
Martin Ross
Mary Janah
Mary Maccallum Sullivan
Mary Thompson
Matilda Peterken
Matt Turley
Matthew Collin
Maura Carty
Maurício Rocha
Max Angle

Melanie Gravel
Melissa Eustace
Meryll Clay
Michael Bowen
Michael Crick
Michael Gulston
Michael Harvey
Michael Hughson
Michael Rozdoba
Michel Puech
Michelle Harris
Michelle Tafur
Michelle Weisstuch
Mike Eames
Miriam Wells
Molly Garris
Nadia Sussman
Nanci Oddone
Naomi Ihara
Naomi Slijkhuis
Nathan Highton
Nathanial Matthews
Neil Boyd
Netta Cartwright
Nina Wallerstein
Oliver Davis
Oscar Salgado
Otavio Cury
Pablo Gonzalez
Pat Goodacre
Patrick Alley
Patrick Ashworth
Patrick Driscall
Paul Carlson
Paul Durham
Paul Edwards
Paul Hanson
Paul Manning
Paul Sherwood
Paula Azzopardi
Paulette Constable
Penny and Mike Derbyshire
Penny Lindner
Peter Casey
Peter Frankopan
Peter Moser
Peter Rigg
Phillip Bleazey
Phillip Elliott
Phoebe Weseley
RA Ryan
Rachel Fischoff
Raymond Roker
Rebecca White
Renato Schermann Ximenes D
Richard Benson
Richard Shapiro
Richard Watkins

Acknowledgements

Rita Schwarzer
Robert Amerongen
Robert Del Naja
Robert of Etruria Cochrane
Robert Williams
Robin Hanbury-Tenison
Robin Roberts
Rosalyn Sparrow
Rose Palmer
Rosemary Collins
Dr Rosemary Jones
Rosie Chandler
Rosie Farr
Ross Anderson
Ruth Dalton
Ruth Morris
Sally Ourieff
Sam Cowie
Sam Stewart
Sandra Staplehurst
Sarah Darwin
Sarah E Robbins
Sarah Gilbert
Sarah O'Sullivan
Seamus M Kirkpatrick
Seana De Carne
Sheena D Rossiter
Shelagh Green
Sian Phillips

Sigrid Houston
Simon La Frenais
Sophie Brown
Sophie Comninos
Stefano Cremonesi
Stephanie Deroo
Stephanie Goodacre
Stephanie Nolen
Stephen Eisenhammer
Stephen Kellett
Steve Gibbons
Steve White
Stuart Grudgings
Sue Gill
Sue Thomas
Sumit Tiwari
Surya Hope
Susan Arisman
Susan Clopton
Susan Lambert
Susan Moreira
Susanna Rustin
SW Lam
Sylvain Machefert
Terry Hughes
Timothy Morris
Tom Blickman
Tom Kissock
Tosca Tindall

Tracey Duncombe
Tracy Thompson
Tyler Bridges
Vanessa Buckley
Veronica Higginson
Vio R
Will Hargreaves
William Castle
William Jordan
William Milliken
William Schomberg and
142 anonymous donors